HOW THE SHORTAGE OF SKILLED MECHANICS IS BEING OVERCOME BY TRAINING THE UNSKILLED

First published 1918
This edition published by Lector House in 2024

ISBN: 978-93-6138-047-1

Lector House LLP
Registered Office: H. No. 96, Block C, Tomar Colony,
Burari, Delhi – 110084, India
info@lectorhouse.com
www.lectorhouse.com

HOW THE SHORTAGE OF SKILLED MECHANICS IS BEING OVERCOME BY TRAINING THE UNSKILLED

COUNCIL OF NATIONAL DEFENSE

2024

LECTOR HOUSE LLP

HOW THE SHORTAGE OF SKILLED MECHANICS IS BEING OVERCOME BY TRAINING THE UNSKILLED

BY

COUNCIL OF NATIONAL DEFENSE

CONTENTS

SECTION ON INDUSTRIAL TRAINING FOR THE WAR EMERGENCY

National Committee

REPRESENTING LABOR:

FRANK DUFFY, General Secretary, United Brotherhood of Carpenters and Joiners of America, Indianapolis, Ind.

HUGH FRAYNE, General Organizer, American Federation of Labor, 706 Council of National Defense Bldg., Washington, D. C.

[1]JOHN GOLDEN, President, United Textile Workers of America, 86-87 Bible House, New York.

GRANT HAMILTON, American Federation of Labor, Washington, D. C.

ARTHUR E. HOLDER, Member Federal Board of Vocational Education, Ouray Building, Washington, D. C.

MISS FLORENCE C. THORNE, American Federation of Labor, Washington, D. C.

CHARLES H. WINSLOW, Federal Board of Vocational Education, Washington, D. C.

REPRESENTING EMPLOYERS:

FREDERICK A. GEIER, President, Cincinnati Milling Machine Company, Cincinnati, Ohio.

HENRY M. LELAND, President, Lincoln Motor Company, Detroit, Mich.

C. E. MICHAEL, President, Virginia Bridge & Iron Company, Roanoke, Va.

[2]PERCY S. STRAUS, R. H. Macy & Co., 1317 Broadway, New York.

[3]H. E. MILES, Formerly President, Wisconsin State Board of Vocational Education, Chairman, Com. on Industrial Education, National Association of Manufacturers, Racine, Wis.

C. U. CARPENTER, Works Manager, Dayton Recording & Computing Machine Company, Dayton, Ohio.

G. B. DUFFIELD (Chairman, Michigan Branch Committee), Detroit Lubricator Company, Detroit, Mich.

[1] Executive Committee.
[2] Executive Committee.
[3] Chairman.

REPRESENTING EDUCATION AND WELFARE:

S. W. ASHE (Chairman, New England Branch Committee), Chairman, Education and Welfare Department, General Electric Co., Pittsfield, Mass.

JOHN C. FRAZEE (Chairman, Pennsylvania Branch Committee), Member of State Committee of Public Safety, 704 Finance Bldg., Philadelphia, Pa.

[4]C. R. DOOLEY, Educational Department, Westinghouse Electric & Manufacturing Co., East Pittsburgh, Pa.

R. O. SMALL, Deputy Commissioner of Education, State House, Boston, Mass.

DR. CHARLES McCARTHY, Ph. D., Chief, Reference Library, Madison, Wis.

ALVIN E. DODD, National Society for the Promotion of Industrial Education, 140 West 42d Street, New York City.

Training for War Industries, which was heretofore developed under the SECTION ON INDUSTRIAL TRAINING of the Council of National Defense, has been taken over by the

TRAINING AND DILUTION SERVICE
DEPARTMENT OF LABOR
618 17TH ST., N. W.
WASHINGTON, D. C.

to whom all correspondence should be addressed.

[4] Executive Committee.

INTENSIVE TRAINING OF UNSKILLED WORKERS AS A MEANS OF OVERCOMING LABOR SHORTAGE

The Committee on Labor, Advisory Commission of the Council of National Defense, of which Samuel Gompers is chairman, authorizes the following from the Official Bulletin of August 14:

The grave situation of shortage of labor (it now being estimated that there is a shortage of 250,000 skilled workers[5]) is being met by a new quick method of training operatives. All over the country, day by day, one factory after another falls into line and puts in a training department to train its own people—the same sort of quick, intensive-training plan to meet the same sort of situation which the regulations in France prescribe for every manufacturer employing 300 people or more, and the English ministry of munitions requires in its contracts for materials. And the situation must be met in greater degree and substantially all factories must train their workers if the 750,000 new skilled workers which the country needs by January 1 are skilled and efficient and standing at their job by that time.

PROVED BY THE FACTORIES

To-day 100[6] important factories making war orders are proving that it is possible to train their own men. They do not assume to teach a worker a whole trade in the brief time available. They do teach him by the methods of the training department how to master one process or one machine in a few weeks or a few days. These 100 factories are spending, or preparing to spend, at the rate of $1,500,000 solely in this business of intensive training of new workers. This training investment is not an expense, as the training is immediately upon production and the product from the training room is expected to equal that in the factory. All the training departments mentioned are on a production basis at all times, with speed and accuracy as the watchword.

DECLARATION OF POLICY

One year ago that section of the committee on labor of the Council of National Defense which has been instrumental in developing the training department or vestibule schools above noted recorded the following as its declaration of policy:

"The Section on Industrial Training for the War Emergency is concerned with industrial training only as a war measure. It is not concerned with vocational ed-

[5] Note: September 30, now estimated 500,000.

[6] Note: September 30, now 200.

ucation in general. In all cases in the existing crisis shortage of labor must be met first by training operatives from allied trades who are unemployed and by advancing operatives of ability from lower to higher positions in the occupation itself. For instance, apprentices should be advanced rather than outsiders. It is possible that many sewing women will be without work, and many men in the building trades. For all such, new and fitting places must be developed where possible. Non-wage earners must not be trained to take places for which unemployed wage earners may reasonably be trained."

At the same time the section on industrial training stated the following to be its plan and scope:

1. Increased use of the public vocational schools through the co-operation of local manufacturers. This is being done very fortunately in Worcester, Bridgeport and some other cities.

2. Introduce new workers, men and women, into industry through these schools.

3. Arrange for the training of present mechanics and others in existing workrooms in connection with regular production, and by more scientific procedure than heretofore.

4. As of particular importance, act as a clearing house, that the judgment and experience, good and bad, in each locality may be available to all.

The section on industrial training, a part of the welfare division of the committee on labor, is composed of one-third representatives of labor, one-third employers and one-third experts in factory training. State committees similarly organized have been developed where war products are being made. There are at present nine associate branch committees of the section on industrial training, which are Illinois, Indiana, Michigan, New England, New Jersey, New York, Ohio, Pennsylvania and Wisconsin.

KEEPING CHECK ON TRAINING

The section recommends three checks on the factory training department, to be made daily by report:

1. How many operatives are sent into the factory? (If this were the only test they might be sent in too fast and only partly trained.)

2. Cost, net, after crediting production which should equal the shop average.

3. Wastage.—There should be none. There should be 100 per cent. Government inspection.

These training departments, because of the thorough training given, have yielded from 10 to 40 per cent. increase in production, both for men and women, and the labor turnover has been reduced materially thereby.

Great care has been taken to advocate that unemployed men be adapted and trained in new trades for the period of the war and that unskilled men be educated wherever possible before resorting to the employment of women.

POSSIBILITIES OF DEVELOPMENT

As an indication of the possibilities of this development, the experience of the State of New Jersey co-operating with this section may be cited. In order to overcome possible objection which labor might have to the introduction of emergency training a program was agreed on, after a series of conferences, which was heartily approved by all the employers and by the representatives of the employees. Some of the clauses of this agreement are as follows:

"All skilled labor available within the surrounding territory should be brought into the essential war industries before it is unduly diluted by the introduction of unskilled labor. When such dilution is necessary, and in the opinion of the committee that time has already arrived, the more skilled activities should be supplied by training those already at work and successful in handling the lesser skilled activities of the same general type. The lesser skilled activities should be supplied by training those already skilled in non-essential activities and not engaged at present in essential war industries, and who, because of such skill, are peculiarly capable of quickly learning the rudiments of the new activities.

"Exploitation of labor and reduction of wages through dilution for war purposes is to be avoided, and to this end persons brought into an essential industry, or promoted from one grade of work to another, are to be paid the prevailing rate of wages for the class of work for which they have been trained, after a training period of reasonable duration.

"Dilution of labor by the employment of women when necessary is recommended, provided women receive wages equal to men for the class of work performed by them, and provided the working conditions surrounding their activities are carefully controlled for their comfort and well-being."

ELIMINATING HOUSING PROBLEM

One interesting result of training resident unemployed is the practical elimination of the housing problem in certain instances.

This is exemplified in the city of Detroit, where it is estimated that 50,000 additional mechanics will be needed before the end of the year. If those now engaged in the war plants could be advanced to more skilled positions, and their places be filled by present residents of Detroit engaged in non-essential or unskilled industries, or those not now at work, the need for housing of the 50,000 mechanics with their families could be, if not entirely, at least, in part, eliminated.

All who have tried these intensive methods of training are happily surprised at the shortness of time required to make skilled operatives for precision work in tool room and factory of men from non-essential trades and of the more intelligent women now entering industry for the war.

ACTIVITIES OF THE CHAIRMAN

To stimulate effort and arouse interest in training the idle and potential workers in each community, as well as to facilitate the up-grading of the old operatives, the sectional chairman has traveled from one manufacturing center to another for the past 12 months, addressing leading metal, machine-tool, and other manufacturers' associations. He has also actually assisted in the establishment of vestibule Training Departments in the plants.

Both single shops and great industrial communities are acting upon the advice of this section, which will furnish experts for investigation and planning upon request.

PRATT INSTITUTE'S NATIONAL SERVICE COURSES IN MACHINE WORK

As a contribution to industrial training for the war emergency, Pratt Institute is conducting day and evening courses in Machine Work, which have been especially organized to serve the present need for increased productive efficiency in this country's machine shops. These courses are designed to aid ambitious machine shop workers of limited development, including machine operators, bench hands and machinists' helpers, who wish to extend and broaden their practical ability as a means to personal advancement in the trade, increased earning power and fuller service to production. Pratt Institute's Machine Shop has been continuously employed to maximum capacity for this instruction since the entry of the United States into the war.

The Machine Work comprises six graded courses, each of which requires for its completion six weeks, if taken as a full-time day course, seven hours per day, or if taken as an evening course, twenty-four weeks, three evenings per week, two hours per evening. Day students register and pay tuition for six weeks, and evening students for twelve weeks. New classes are started at frequent intervals. A student may start in any course for which he is qualified, and may enroll for additional courses, either consecutively, or at some later time, if he finds it desirable to withdraw temporarily. Students are permitted also to transfer at any time from the day to the evening course or vice versa, with full credit for work already performed.

The instruction is adapted to the individual. Men capable of following directions without excessive damage to material or equipment are put on productive work. Men who have not reached this degree of efficiency are commonly assigned exercises. About 75 per cent. of the work is productive. Production is introduced as a means to greater efficiency in training.

Courses in Wooden Boat and Shipbuilding, Marine Engineering, Gasoline Engine Maintenance and Operation, Machine Drawing and Design, Ship Drafting, Chemical Laboratory Practice, in addition to an extensive list of day and evening trade and technical courses, all of which are of special service in the war emergency, are being conducted.

(Signed) Samuel S. Edmands.

Pratt Institute's Machine Shop has been employed to capacity since our entry into the war.

Pratt Institute. Note the "older men."

Pratt Institute (Brooklyn) is conducting day and evening courses in machine work.

Pratt Institute. Ambitious machine shop workers go out from here to give fuller service in production.

BOARDMAN TRADE SCHOOL

The Boardman Apprentice Shop, under which name New Haven, Conn., operates a Trade School, is doing its share toward meeting the shortage of skilled and semi-skilled help and plans are being made to further this work.

The "Shop" teaches many trades under actual trade conditions, but as the most pressing need is for machine workers this trade only will be considered in this article even to the exclusion of the drafting department, second in importance, and results to the machine department.

Primarily this Trade School is operated to teach boys, but the evening continuation classes have grown in importance year by year until they have reached the present high standard of efficiency.

The machine department trains fifty boys in the day course and under normal conditions the boy graduates after 4,800 hours of study, seventy-five per cent. being trade practice and twenty-five per cent. academic study. At present many boys leave before the completion of their course to enter local munition factories. These boys are in great demand and even after a few months of training are found extremely useful in those factories.

The boys who complete their studies and receive their diplomas are largely sought for tool room work.

Thus the school is supplying more trained hands than the number of boys and length of course would indicate and that is but part of the story. These boys work eight hours a day, forty-four hours a week, fifty weeks a year, and produce real machinery practically all of which goes into the munition plants.

One lot of forty-five Horizontal Tappers was built and boxed and ready for shipment to Glasgow for use on British munitions long before cargo space was available.

The boys build two sizes of screw slotting machines, two sizes of horizontal tapping machines, lathes, slide rests, drill press vises and hundreds of small cutters.

They have built and shipped about six hundred machines, not including slide rests and vises.

The screw slotters and tapping machines are of a type in great demand for munition factories, being particularly serviceable for use on fuse parts, small arms and government hardware.

Thus, the school, while following its basic plan, is supplying the country's vital needs in training boys and at the same time making an essential product.

In addition, further use of the equipment is secured by the operation of night continuation classes for twenty-five weeks in the year. The classes are operated six nights per week and Saturday afternoon with instructors taken from the local factories under one of the regular day force.

Men in all stages of experience, ambitious apprentices, unskilled clerks, drivers, porters, etc., who wish to enter the local munition factories come to these classes.

As an instance of extremes we may take the case of a painter of sixty who entered the Marlin Rockwell Corporation on machine work after two seasons of study; and the case of an experienced toolmaker taking advantage of the equipment to learn some new operations so as to fit himself for a higher grade of work. Both men made good. Machinists take the continuation course so that they may qualify as toolmakers. Four classes of fifty men—making a total of two hundred men—are taught in the night classes.

The results have been so satisfactory that these classes will be continued and if the demand warrants women will be given instruction upon specified evenings. The school management believes that the day is coming when women will receive a far greater share of trade instruction.

Plans are in operation to increase the efficiency of the school by teaching special classes. For instance, a class in the use of measuring instruments and gauges would prove valuable and reduce, in a great measure, the time taken in training the unskilled men and women taking up factory work. Large numbers can be handled in such courses.

The school is prepared to take crippled and disabled resident soldiers in any of its trade courses, night or day, when the demand comes.

Since this article was prepared, orders for fifty No. 1 screw slotters and twelve No. 1 vertical tapping machines have been booked at this school.

This order of twelve tapping machines will go directly into an optical factory for use on government supplies.

FRANK R. LAWRENCE,
Acting Director.

September 16, 1918.

BOARDMAN APPRENTICE SCHOOL, NEW HAVEN.

Milling the "tailstock" on a motor-driven vertical milling machine. A natural mechanic (one out of every fifteen that apply can be classed as such), 14 years old. Has been an apprentice about four months. Is doing work usually done by boys of eighteen months' experience.

BOARDMAN APPRENTICE SCHOOL, NEW HAVEN.

Boring is an advanced branch of the machine trade, and requires great skill to successfully complete an accurate piece of work.

A boy must complete 4,000 hours before he is advanced to this operation, and not then unless we consider him competent to do this accurate work.

The "head" and "tail" of this machine must "line" to .001 of an inch in 18 inches, and therefore must be bored until all the "spring" is out of the boring bar.

This boy, age 15, is making a measurement with a spring caliper to ascertain proper size before reaming.

BOARDMAN APPRENTICE SCHOOL, NEW HAVEN.

Scraping beds—a difficult art. Notice the standard Brown & Sharpe surface plate at the left. The surface of these beds must show an 85 per cent bearing, the tailstocks scraped to fit the same. These boys are about 14½ years old, and have served six months' apprenticeship.

BOARDMAN APPRENTICE SCHOOL, NEW HAVEN.

Planing "head." It is one of the advanced operations and requires much care in machining. The slot shown must be absolutely in line with the boxes, and they are tested with an aligning bar after planing. This boy is 16 years old, and will graduate in about two months.

DAYTON INDUSTRIAL INSTITUTE

Dayton, Ohio

The Dayton Industrial Institute was established to replace the vestibule schools in the following plants: The Dayton Engineering Laboratories Co., the Domestic Engineering Co., the Dayton Metal Products Co., and the Dayton Wright Airplane Co.

Although the school has been fostered by the above companies, any manufacturer in Dayton is at liberty to send students to the school under certain regulations.

The Directors of the several companies thought it advisable to segregate the training school from the plants, and combine the school under one directing head. Most gratifying results have been obtained since the opening of the school January 1, 1918.

During the past seven months over 500 persons, from all walks of life, have been trained for factory work. About 200 of this number were women.

In addition to the large number of people trained, over 100,000 pieces of commercial product have been manufactured, which passed the most rigid factory inspection.

A large percentage of the work has been parts of war products, such as detonator bodies, Liberty motor ignition parts, inspection gauges for war materials, crank shafts, cam shafts, motor truck parts, etc., as well as airplane parts, consisting of ribs, fins, stabilizers, wheel covers, etc., for the DeHaviland fighting plane.

Would it not be advisable for manufacturers in some communities, and especially for small manufacturers, to pool their interests in regard to industrial training, in order that the schools may be a peace time asset, as well as to satisfy a war time emergency?

WRIGHT-MARTIN AIRCRAFT CORPORATION

The Wright-Martin Aircraft Corporation has two shop training departments, one at the Long Island City plant and the other at the plant in New Brunswick, N. J. Both plants are engaged in the manufacture of a high-grade aeroplane engine.

Each training department occupies about 10,000 square feet of floor space in buildings separate from the factories. They are equipped with modern machinery tooled up for production, the machines and equipment being of the same type as that in the factory proper.

The primary purpose of these departments is to train women, also men, for the needs of the factory, upon production work, assembly, inspection, shop clerk work and tool crib tending. It is the aim of the Instruction Department to train the learner to do her work habitually correct, both as to quantity and quality. With this in mind the training rooms are miniature factories equipped with lathes, automatic screw machines, hand screw machines, J. & L. Turret Lathes, hand millers, plain millers, sensitive drills, upright drills, radial drills, plain grinders, internal grinders and bench equipment. Jigs and fixtures and the operation tool equipment necessary for the production work are used on these machines, that those being trained may become entirely familiar with the tools they are required to handle when they go into the shop. The machines and the operation tool equipment were selected after the complete layout of the main operations had been carefully gone over and the operations that women could perform were listed.

Considerable care has been used in selecting the learners and the class of women under training are of good type. A number of these have worked in shops before; others have been in offices. Our experience has been that our most successful women are those who have had to work for their living, either in shops or offices, up to the time they entered our employ. The range of ages preferred varies from 21 to 35. The first few hundred women selected were about 35 years of age, the maturity and judgment going with such an age being of value in stabilizing later conditions when hundreds more, of a more general sort, will be employed. Many of the women are mothers, wives, or relatives of those at the front.

The training is given upon actual factory production in the manufacture of parts that enter into the construction of the motor. Such manufacturing furnishes an excellent medium for instruction in the various branches above mentioned, and holds both the learner and the instructor up to the factory requirements. The standards of the factory are the standards of the training department both as to quantity and quality. Work is routed to the Instruction Department according to the regular shop forms and the finished product is transferred from the Instruction Department in the same manner as work is transferred within the factory, the Instruction Department receiving credit for what it does.

So far, women are being trained at a rate equal to the demands of the factory, which is fast approaching 120 per week, this being the approximate weekly training capacity of the training department. In some branches of work it takes four days to train, in others ten days, the length of time varying according to the time it takes the learner to reach the average hourly production.

Records of production, while under training, are plotted on cross section paper and when the learner has reached the average hourly production she is declared trained in that particular line. In some cases women after two or three days' training have done 25 per cent. higher than the average hourly production stated for the job. Records of salvage are also kept as a check on such training, and every effort is made to combine a steep production curve and a minimum salvage curve with good training. The salvage records of the learners are remarkably low, some weeks averaging much less than 1 per cent., the highest being less than that in the factory itself.

After the women have been transferred to the factory their progress is kept track of for at least one week, to make sure that they are following the instructions given in the training room. Those who do not make good after training are assigned back to the training room for further instruction, or for such disposition as the chief instructor may see fit.

Regularly, during the week, the learners are given general lectures, talks and instructions on matters that relate to their training, that they may be more generally fitted for the line of work into which they are to be transferred.

The results obtained have been very interesting. The women are very enthusiastic and the foremen are highly pleased. In one branch the foreman advised, when asked how things were going: "You can give me thirty more women right away; they are all right," In another branch a foreman advised that he would not exchange a good share of the women in his department for an equal number of the best men he had on his floor.

The instruction in each training room is given by four male instructors, with women assistants; the women assistants having been selected from the best of those who have been trained within the department. The men instructors are all first class mechanics, especially capable on production work and teaching.

The instructors are regularly interviewed from time to time, and the work they are doing is carefully reviewed, with the purpose of building them up as efficient instructors.

As a part of the training program at New Brunswick an evening school is conducted for the men in the company's employ. The instruction work given consists of technical studies related to the mechanical trades, and includes blue-print reading, shop drawing, shop mathematics and clerical work in its different branches. This is used in conjunction with a promotion program, whereby men who are capable are promoted into various openings as they occur, requiring more skill of the same sort they already have.

(Signed) James F. Johnson,
Chief Instructor.

Learning to operate the radial drill presses Wright-Martin Aircraft Corp.

Being trained upon an engine lathe to accuracy of one-thousandth.

Training on the job in shop. Wright-Martin Aircraft Corp.

Being trained upon a James & Lamson machine.
Wright-Martin Aircraft Corp.

WINCHESTER REPEATING ARMS COMPANY

New Haven, Conn.

Manifestly, all large employers of labor must train their help to a greater or less degree. It has always been the practice of the Winchester Company to do a great deal of this training, but until the past few years it had all been done in the shops or office where the candidate was to work. For many years it has been the practice to take intelligent boys and young men in the office with the idea of continually promoting them to more responsible positions after they became competent and the positions developed where they might be used. Our graduate apprentices, adjusters and tool setters have been so well trained that they have always been sought for by competitive concerns. Since the war, and especially during the past six or eight months, it has become increasingly necessary to develop the training of employees to the greatest extent. Where formerly help might be obtained who had, at least, some knowledge of machines and shop practice, *it is now necessary to take help who have absolutely no knowledge of factory work and teach them to become skilled*. In addition, *this must be done in the least possible time.*

To take care of this condition, we have developed in addition to our regular apprentice course, a training course for administrative and executive positions, an Office School, a Gun Department Adjuster's Shop, a Cartridge Department Training Shop and Tool Department Training Shop.

The regular apprentice course is designed to give complete training for machinists, gauge makers, tool and fixture makers, etc. The course which ordinarily requires three years to complete has recently been shortened so that some of the boys are graduated in two years. It is the desire to give thorough instruction in the practical work mentioned above. The average enrollment is approximately 90.

The training course for administrative and executive positions is designed to cover briefly such shop and office practice as will give the broadest general knowledge that is likely to be required of those in the more important positions in the administrative organization. This class consists of only about a dozen men who are picked with all possible care.

OFFICE SCHOOL

The Office School under competent instructors consists of a group of clerks who do the complete work of making up pay roll, labor distribution cards, etc., for a number of factory shops which were chosen because of their having respective classes of work. Prospective clerks are taken into this school, trained on pay roll work, transferred to the Central Pay Roll Division as help is needed. In addition to this school for pay roll clerks, model shop offices are being established in each of the major departments where correct methods will be carefully taught. It is our aim to have all shop clerks pass through one of these model shop offices to receive their training, as this will insure standard methods and proper following of procedures. These model shop offices will also act as reservoirs on which to draw to supply vacancies caused by absences among the regular shop office staff.

GUN DEPARTMENT ADJUSTER'S SHOP

The shop for training adjusters for the Gun Department is designed to teach men how to adjust the type of machines to which they are to be assigned for all classes of work which may be run upon them. This training should ordinarily take *from two to four* weeks, but many times it has seemed desirable to graduate candidates more quickly than this; if they are bright and intelligent it has proven satisfactory to do so. No attempt has been made to train operators in the Gun Department except in the shops where they are to work. Most of this work can be learned in a *very few* days under the instruction of a well trained adjuster.

CARTRIDGE TRAINING SHOP

The Cartridge Training Shop was designed to train adjusters and tool setters, and some operators. Due to the great number of new employees, it has been impossible to have all tool setters and adjusters pass through this school, but it is hoped that in the near future we may be able to train an increasing percentage of them here. Such men as have received this training have shown beyond doubt that they have been much benefited by it, and that it is most desirable to expand it to include as many of this class of workmen as possible. Some operators have been trained in the Cartridge Training School with excellent results, but most of the work is relatively simple and the training has been satisfactorily done in the shop to which the new employees have been assigned.

TOOL DEPARTMENT TRAINING SHOP

A new Tool Department Training Shop is just being started for the purpose of training operators on lathes, milling machines, planers, grinders, etc. At first we tried to train this kind of help in the Apprentice Shop, but because of the fact that the kind of instruction was so vastly different, it has been decided impractical. In the one case, we wish to give very complete, broad instructions, and in the other, the desire is to train for one kind of work only in the least possible time.

MANUFACTURING TOOL SHOPS

We have two shops working night and day offering facilities for the training of unskilled people in certain lines of tool work. These people are taken in totally without experience, and placed under the tutelage of our best mechanics, and trained quickly as specialists. Within a remarkably short time they are capable of producing all sorts of tools used in the production of guns and ammunition, thus relieving the general tool shop of a great volume of work which would otherwise require the services of skilled tool makers. From the forces in these shops we are able to recruit the more advanced men for gauge, jig and fixture work. We consider these shops one of the best examples possible of training upon a productive basis.

Aside from the regular training courses as outlined, there are many instructors throughout the shops whose duty it is to explain the best ways of doing the various tasks to which employees are assigned. It has been our intent throughout, in choosing instructors, to select those who are real teachers, having the necessary patience and human understanding required to successfully do work of this kind.

The regular source through which we train toolmakers is the Apprentice Shop. This shop has an enrollment of 125 boys, who are trained in a three-year course to become expert all-round mechanics.

In addition to this, two of our largest shops, of 250 men each, have during the past year and a half conducted special training courses for green men on the more elementary work of toolmaking. The purpose of these courses is to train men who have had no mechanical experience to the point where they can be used to free more expert men from all routine and simple work. These men are started on a lathe grinder or milling machine. Those men who show special aptitude are taught how to use two or more machines, while the rest are trained to operate only a single type. *By means of our special course of instruction it is possible to train new men in a period of a week to four weeks,* depending on the man and on the kind of work for which he is being trained. During the last year and a half over 500 men have been trained in this way. Recently 60 men were trained in one month. The best of these are advanced to more difficult work, and some of them even become third class toolmakers.

In addition to this work, we have been training girls since last March to do machine work. These girls have hitherto been sent directly to the Apprentice Shop and there given a week's training on a lathe or a milling machine on repetition work. After this they were transferred to one of the regular shops where they have done extremely good work.

In addition to the training of expert toolmakers and mechanics for the Tool Department, we are also training mechanics for the Gun and Cartridge Departments. Each of these departments has a school or training department containing representative machines and in these schools men are given a course of instruction lasting from one to two months. This prepares them to go out into the shops and take care of a group of machines, keeping them in repair, supplied with proper tools, and generally in good running order.

Third, since the above was still not sufficient to fill our requirements, it was decided to start a regular training shop in which to train machinists and toolmakers, as well as gauge-makers. These men were to be trained in the use of three or four machines; the lathe, miller, planer, shaper and grinder. They were to do the more simple work on these machines, but still, work that was not repetition. In just three weeks after the plan had been accepted, the space was secured, and the equipment of 30 machines, tools, and everything which goes with a complete shop, including overhead shafting, was installed. Moreover, a complete set of drawings representing 40 typical operations was made and blue-printed, and these will serve as a plan of instruction. On Monday morning, three weeks after the plan was approved, the shop, with a complete personnel of foremen and instructors, began operation. *It is planned to turn men out in from three to five weeks.*

In all our work the emphasis is on production. Training wherever possible is given on actual production work. This not only makes possible training of a very practical nature, but also helps to lessen the cost of the instruction.

(Signed) L. O. PETHICK,
Personnel Superintendent.

NEWLY TRAINED OPERATOR

—TOOL TRAINING SHOP—

GIRL LATHE OPERATOR

—CUTTING TOOL SHOP—

BROWN & SHARPE MANUFACTURING COMPANY

Providence, R. I.

In our own experience, without doubt, much more attention has been given by foremen and fellow-workmen to the supervision of women's work than has been given to the average male employee in the past, the assumption being that a woman, having less mechanical background and intuition than a man, required more training and more specific instructions. This has been the reason advanced by some foremen in explaining why women were doing better work and had "broken in" more quickly than men, and they have added, "If we had given the same kind of attention to each new man employed, he would have done just as well as the girl"; this is, after all, an admission on the part of the foreman that he had not in the past helped all he could, and an indirect compliment to the girl having much significance. It may be noted, however, that at the time when such comparisons were made the average man who could be secured was of an unsatisfactory and irresponsible class, as so few trained or competent men were available for positions in the industries, while, on the other hand, in hiring girls a selection from a large number of applicants could be made, so that it was possible to obtain a much better average having the qualities to make successful workers.

Experience has shown that there are advantages in having both men and women in the same department, as it tends to hold the same standard of workmanship and speed for women as for men, while it is believed that having a separate department for women may establish a separate and lower standard, the tendency being to make more allowance for women because of sex. The results seem to show that it is not at all necessary that separate standards should be established and that in some lines of work even more can be expected of women than of men because of their nimble fingers and quickness of motion. As to questions of discipline, where the two sexes are employed in the same work-room, little or no difficulty is experienced under capable foremanship.

Actual results have proved that the fears in the minds of some that there would be opposition on the part of foremen and workmen to the employment of women in the shop were ungrounded. A foreman remarked to a visitor: "See that girl working beside the man assembling speed indicators? She is working with him so as to learn all the requirements, and he knows that she is to have his job as soon as she has become sufficiently proficient, but he is helping her in every way possible. Of course, we shall find other work for the man; and often, with the present shortage of help, such a change of work can be in the line of promotion." This illustrates the spirit which is practically universal throughout the shop, and which has been an important factor in bringing about the success of the plan.

While the money question—the earning power—is uppermost in the minds of the majority, many of the women show also a distinct ambition to equal or excel men in the work they do. Soon after the employment of women was begun in the gear department, a girl who was cutting sprockets on a gear-cutting machine became discouraged and said she was afraid she could not make a success of the job. Her foreman was surprised and said to her, "We have not made any complaint as to your work, have we?" "No," she said, "but the man who worked on the night job turned out 105 pieces, while the best I could do was only 85 pieces a day." Her foreman asked if she realized that the man on the night force was working three hours more per day than she was, and after learning this she felt less discouraged with the results she had obtained.

In the gear department where a number of girls have been "broken in" in operating gear-cutting machines, the foreman said that they had taken hold as quickly as the average man, and some of them are doing exceptionally good and intelligent work. This has partly resulted from the girls being thrown as rapidly as possible on their own resources, being taught to set up their machines, working from a blue-print, to measure their work, and do everything that had previously been required of the operator. A criticism has recently been made of some of the departments to the effect that the foremen were giving so much supervision to the women's work that they were not thrown sufficiently on their own resources, and thus were not trained to be responsible for the work in hand. This again speaks well for the women, as showing that there is a growing appreciation of their ability to do more advanced work than had at first been expected.

In inspection work a field has been found for women in which they are making an exceptionally good showing. The chief inspector was asked whether women were learning to read the micrometer caliper. He replied that they learned to read it and read it accurately, in a very short time, and that the work passing through their hands showed much discrimination as to the points criticised. He pointed to a pile of work rejected by one of the women inspectors and said, "I have just had a man go over this work, and he has found that while the work failed to pass inspection for many reasons, they were all good reasons." He said further that in inspecting grinding work he was surprised at the quickness with which some of his women inspectors would pick out batches of work identifying them as coming from particular workmen whose work was known to be above the average. In another department, in inspecting measuring tools, a similar condition was noted by the foreman, and he stated that one of the girl inspectors recently told him that she liked to inspect the work of Mr. Blank, because it required so few rejections. "And," remarked the foreman, "she sized the situation up just right." He also showed the writer the notes attached to a number of tools which had been held out by the woman inspector for corrections, these criticisms showing much discrimination on her part, and as good a degree of judgment as would have been expected from the experienced inspectors who had previously been doing the work. It is thus found that in the class of inspection work where women are employed the standard is not lowered because of their employment.

Already several women are employed in the toolmaking department. One of these employees, who was operating a lathe turning out tool-steel blanks for bits and reamers, doing her own setting up and measuring, evinced enthusiasm for machine shop work, showing, in reply to questions, that her work was opening up a new field in which she took especial interest and she remarked, "No more housework for me," with such feeling that it was evident her interests strongly leaned in a mechanical direction. Girls in the toolmaking department are working on universal milling machines, surface grinders, etc., as well as lathes. Some of the younger girls throughout the works are employed as messengers.

(Signed) L. D. BURLINGAME.

Operating automatic gear-cutting machines. The girls are taught to set up their machines and make all measurements, working from a blue-print. Brown & Sharpe Mfg. Co.

Girls employed on polishing machines. "When they become proficient on polishing, they are given more advanced work at "hand-tooling," etc., on these machines. Brown & Sharpe Mfg. Co.

Fluting reamers, etc., on Universal milling machines in the toolmaking department. Brown & Sharpe Mfg. Co.

Operating hand milling machines in the production of duplicate parts. Brown & Sharpe Mfg. Co.

TOOLROOM ANALYSIS

What does tool room work consist of? It can be separated into two divisions, namely, that which only the skilled toolmaker can do, such as laying out, fitting, assembling and devising special setups. The balance is machine work. Let us consider the latter and see what it consists of.

Machines. The machines used are generally the engine lathe, the horizontal and vertical millers, planers, shaper, plain, flat and universal grinders, drill press and filing machines. Hobbing and bench lathes may be added.

The Lathe. The lathe work usually consists in machining work preparatory to hardening and grinding. This may be roughly divided into turning, chucking and faceplate work. The toolmaker frequently does all of this work. Let us relieve him of this work and use our semi-skilled man to do it for him. Let one man do the turning, another the chucking and another the faceplate work. Where buttons are used the toolmaker sets them at the bench, but the lathe hand can be instructed on the set-up and the use of the warbler and indicator. The same method can be used on the milling machine in boring holes for bushings, etc. If there is not enough turning the same man can do the boring. The whole result is the toolmaker can supervise, or carry through, a number of jobs at the same time. The lathe hand saves time because he has his tools ground, his straps, parallels, etc., ready at hand and he knows his machine. It is the writer's experience that more time is lost by the toolmaker in hunting for these accessories than it frequently takes to do the job.

The Shaper. On the shaper the work can be divided into roughing out and following an irregular line. Let us take a blanking die for an example. The diemaker (who works at the bench) lays out the die on rough stock. Then it is roughed out to this layout. Next the tap-holes and swivel pins are drilled, tapped and reamed. Now the diemaker can set his die up and lay it out more accurately to his turnplate. The better shaper hand now machines the pieces to his lines and all the diemaker has to do is file the clearance on the cutting edge and remove tool marks. In a sectional die he frequently leaves this for the grinder to do. Now the punch, which has been roughed out, is "sheared" and the machine hand machines the surplus stock from it to make it easier for the diemaker. The head block, shoe, knockout plate, sub-press pins, and stripper have been machined and are ready for him. The drill press hand does the drilling and reaming for the pins and springs, also the tapping. By this method of using the semi-skilled machine hands the skilled man can carry on five and more dies at the same time and *not lose any time waiting for machines*.

The Miller. The milling machine work can be divided into flat work, cutting teeth in cutters, reamers, gears, etc., spiral cutting, boring jigs and special outline work.,

Take the first two groups. The semi-skilled operator can be easily trained for this work, as it does not call for more difficult work than the use of the dividing head, and there is always a chart for that. Spiral cutting can be taught, as there is a chart for that. Cutting cams is more difficult but if there is enough of it the operators can be taught to do it.

The boring of drill jigs, and similar work, can be done by an operator because it is laid out beforehand by the toolmaker, or diemaker in case it is a die.

The Drill Press. The drill press work presents the same solution. The skilled tool or diemaker makes the layout and then drills the holes to the layout. Why use the skilled man's time when a lower-priced, and less valued, operator can be used to drill to this same layout?

Grinders. The horizontal, the plain (or flat) and the universal grinding machines have always had specially trained men so we need not consider them here.

Special Machines. The remaining machines in the toolroom are generally special machines with men to operate them. The toolmaker uses, for the most part, only the machines considered in this article.

Training. The question now arises, "Where will these men be trained and who will train them?" I offer this answer. Men on these machines throughout the factory are semi-skilled in their use and are mostly on repetition work. Take the best of them and train them in the Training Department, or in the tool room, and replace them by new men in the factory.

Results. This method results in: First, enabling the skilled tool or diemaker to handle more work than if he had to do all the machine and layout work; second, increasing the output per machine, for it stops the time lost through the machine's being idle and the tools being separated from the machine. In this matter alone it presents a saving, as it calls for only one set of tools per machine, against a set for every man in the room who keeps them in his bench drawer most of the time; third, it eliminates the time lost by the skilled man's waiting around for a particular machine. He is now able to plan one job after another and turn it over to the machine operator and thus devote all his time to work that an unskilled man cannot do.

(Signed) Walter F. Maddison,
Director of Industrial Training.

THE BLANCHARD MACHINE COMPANY

64 State Street, Cambridge, Mass.

As you know, we have been running our Training Department for about five weeks only, hence we are not in position to give you any definite information as to the value of it, etc., but from what we can see it will be undoubtedly a great help to us, because all the unskilled help go to the instructor before being put into the shop. Those who have had some experience are put into the shop, with the instructor to give them detailed information for as long a time as is necessary, and to teach them the important parts of the work in hand. This, as you can imagine, is more difficult in this shop where we do not manufacture large quantities than it would be in a shop where there was a uniform operation, such as there would be on shells, and work of that kind.

We have taken cabinetmakers and taught them to run boring mills; blank book salesmen to assemble units for our SURFACE GRINDERS; shoemakers to assemble units for SURFACE GRINDERS; carpenters to run turret lathes, and plain helpers or sweepers to break in on Surface and Floor Motor Grinders.

We also have a number of women in the shop whom we have taken in without their having any previous experience in machine work, and taught them various operations, such as broaching, bench work, drilling, turning bevel gears, vise work, cutting long threads, and work of a similar nature, and have found them very satisfactory on this class of work.

We have endeavored to teach them the rudiments of this work before putting them on to regular production work, but after they master the first part of it, all the work that is done is on a regular production basis, and we have found in a great many cases that they have been able to reduce the time taken per piece to a very marked degree over what has formerly been taken by men.

I send herewith eight photographs of our operatives that have been broken into skilled work of various kinds throughout our shop, that has previously been done by men skilled in the particular line involved.

We think that some of these are almost remarkable, when we consider what our attitude was two years ago on work of this kind, refusing absolutely to put anyone on who had not been skilled in the particular line involved.

(Signed) WINFIELD W. BLAKEMAN,
Superintendent.

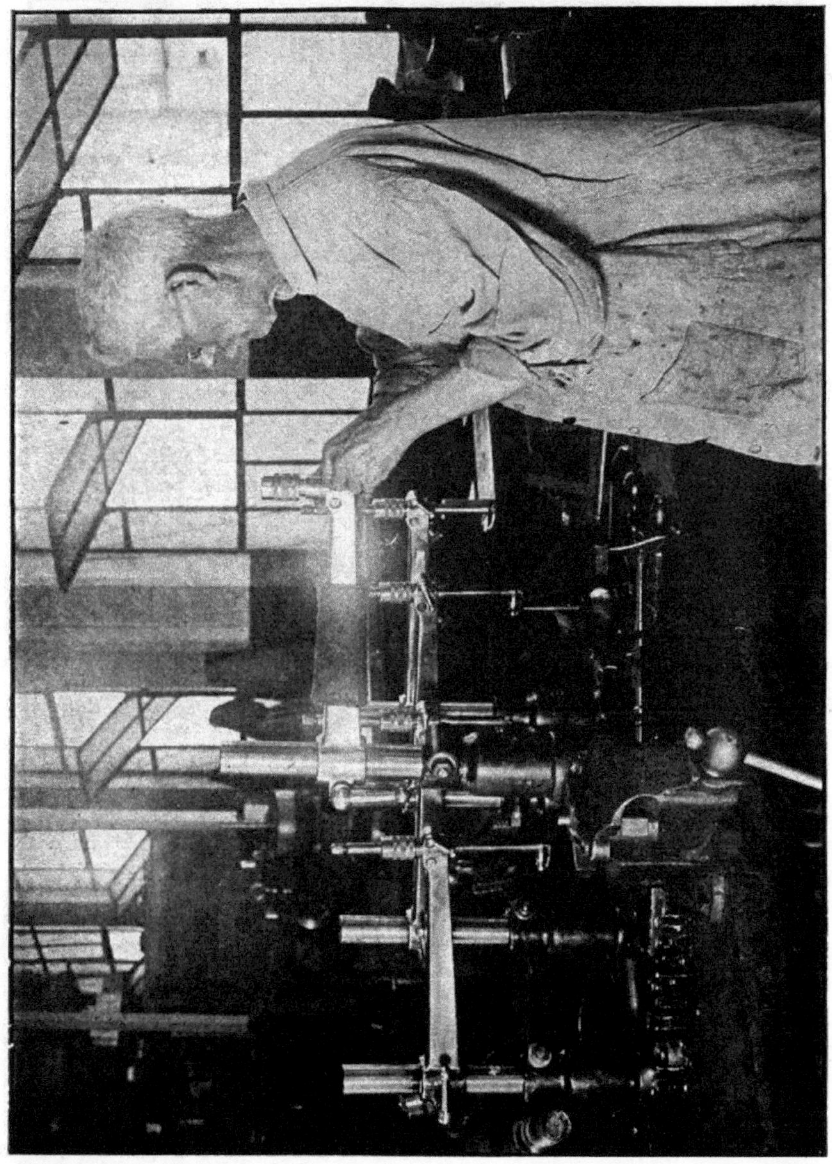

Assembly of our caliper device used in connection with our high-power vertical-surface grinder, for fine measurements on parts being surface ground.

Done by "a man 63 years old, a shoemaker by trade, who has been on this work since June 27, 1918, and has learned in that time to completely assemble these delicate instruments, making the proper adjustments, lapping and doing a quality of work that passes a rigid inspection." Blanchard Machine Co.

Thread-cutting operation on a feed screw for our surface grinder, which is made from a forty-carbon steel, is 26½ inches long and has one-quarter inch pitch acme thread about two-thirds of its length that must be a close fit in a bronze unit.

Done by "a young lady, who has been on this class of work since May 9, 1918, has been able to take these screws from the rough stock, turn them to grinding size and finally finish cutting the thread in a time that is less than was formerly taken by skilled machinists. We think that this is one of the most remarkable jobs done by the women in our shop, as this work requires very careful attention and unusual skill." Blanchard Machine Co.

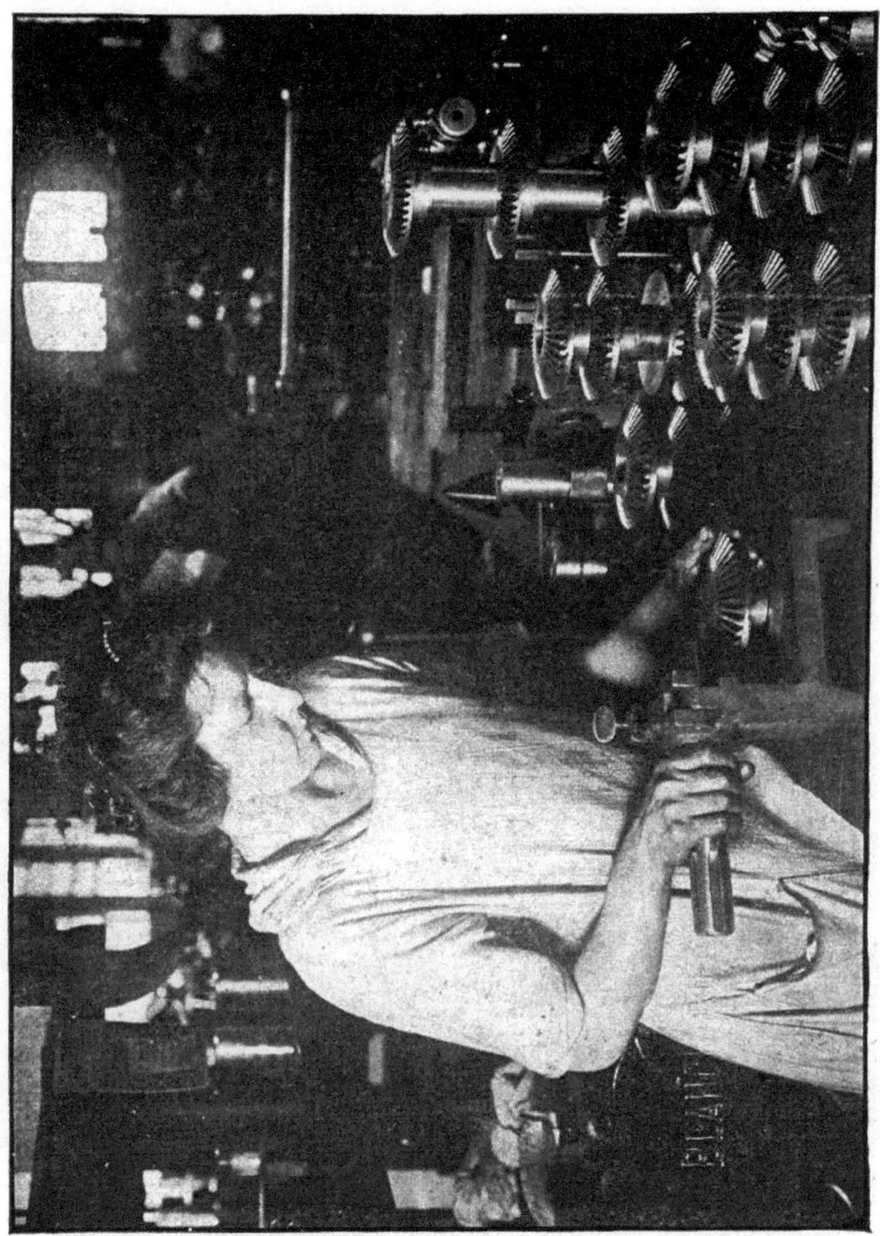

Assistant Inspector. Blanchard Machine Co.

Done by "a young girl of twenty years, who has been assistant inspector since April 29, 1918, and while she does not understand all the technical phrases used in connection with, work of this sort, there is a very large percentage which is merely routine, and if it does not pass the gauges provided she refers it to another man to put on production work."

Machinist. Operation of Turret Lathe. Blanchard Machine Co. Done by "a young man, carpenter by trade (not in draft), having no previous experience on machinery but by keeping a uniform line of work going through this machine, and giving him careful instructions, he is able to almost equal that of a skilled operator. He has been in our employ since January 9, 1918."

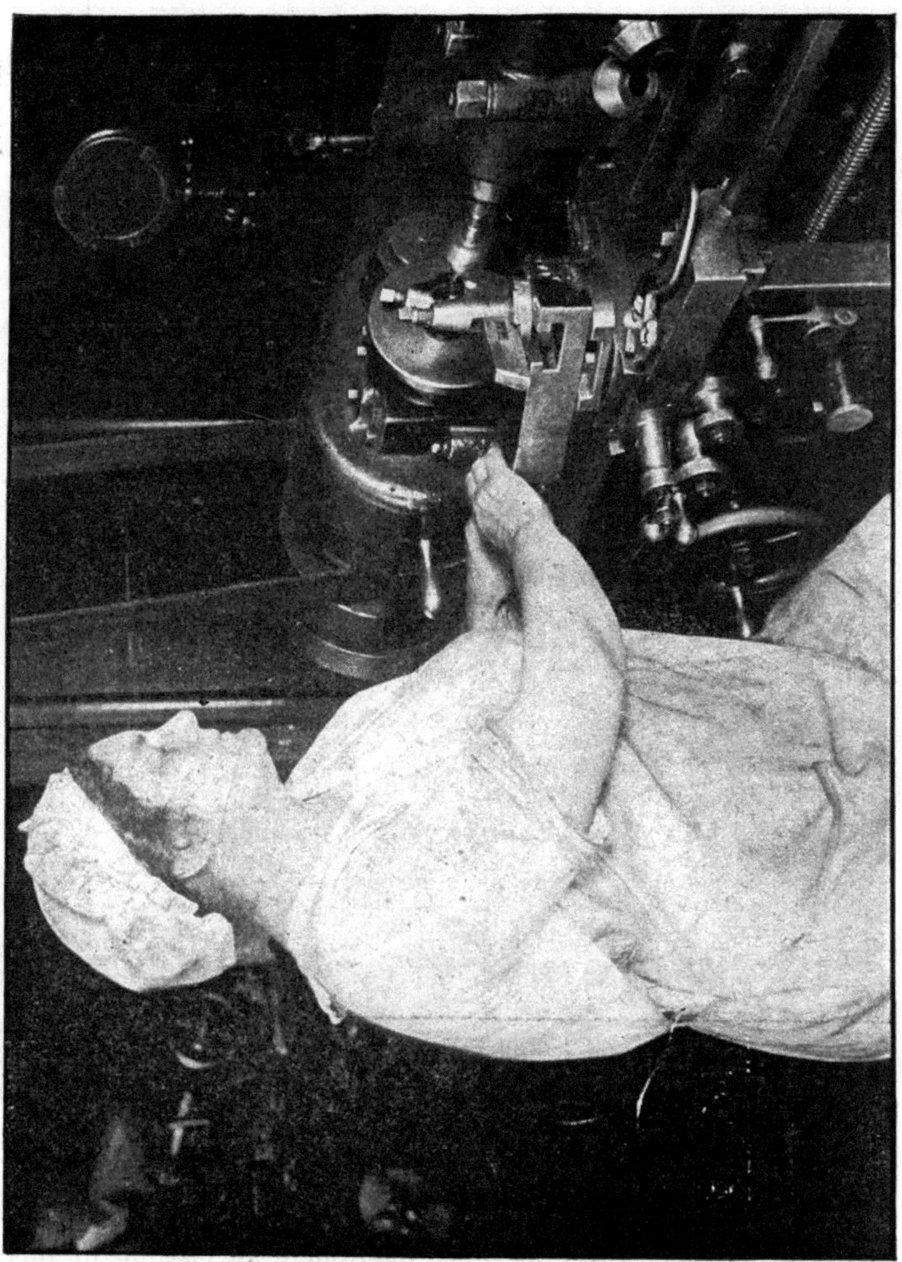

Finish turning of steel bevel gears to accurate dimensions, using compound slide and producing a quality of work that will pass the most critical inspection.

Done by "a woman who had no previous experience on lathes and came to work in June of this year." Blanchard Machine Co.

Graduating of the segment plates used on our high-power vertical surface grinder. It calls for the use of an index head and making every fifth line three-sixteenths of an inch longer than the balance of the lines in the section, using a gauge shown on the front of the machine for this purpose. Blanchard Machine Co.

DONE BY "a young lady who had a small amount of experience assembling on automobile starters before she came to work for us on June 24, 1918. By setting up the machine and not giving her too difficult jobs, she is able to about equal the time taken by the average man."

Jig drilling on a 20-inch upright drill that has previously been done by men. Blanchard Machine Co.

DONE BY "women operatives who are usually able to practically reach the time taken by the men on this work. In a few cases where the men have had a longer period of experience they have been able to improve on the time taken by the women. A great deal of this work calls for drilling and tapping, as well as counterboring and spot facing."

THE AMERICAN SHELL COMPANY

Paterson, N. J.

In reference to our method of producing toolroom outfits most economically and with a large output:

This is chiefly accomplished by anticipating our machine shop requirements or demands far enough in advance so that the majority of work passing through the toolroom can be manufactured in quantities great enough to be produced economically, and also bringing the toolroom work nearer to an actual manufacturing basis. This also eliminates a lot of down or lost time, including the unnecessary losses, such as occur between jobs on work of this nature.

It also decreases the percentage of losses due to spoiled work as workers become more skilled in performing one kind of work rather than a number of different classes of work, which is invariably the case in toolrooms.

We also find this gives us the opportunity to take advantage of present conditions and hire a man who sometimes is inclined to call himself a toolmaker, although he is not an all-around man but is still quite suitable under our conditions of working.

We find, too, that on quantity production in toolroom, we are in a position to effect some very great savings in time and money, as it permits standard jigs and fixtures, such as in ordinary manufacturing; for instance, we at one time machined all our flat tool bits made of high speed steel before hardening and grinding, while our present method is to heat bars of steel in gas furnace and then punch out with die under heavy hammer or press. This method alone in a couple of hours gives us a supply ahead that would ordinarily take one machine or more in continuous operation to produce.

The classes of work in toolroom are also segregated as much as possible; we have one gang for machine repair and reconstruction only. Another gang does all the roughing work on tools and gauges so that the finishing work only on gauges is done by the gauge-makers who are also practically a separate gang. The grinding of tools is under one skilled mechanic who is a working gang boss and leader of men trained to do this work only.

Our tool supply is never permitted to get below the fixed minimum quantity, and record of this condition is always before the toolroom foreman and maintained by his clerk. His clerk is advised hourly on this through the disbursements which he receives from various shop cribs.

We also make a daily record of these same disbursements as against operations and production, thereby keeping in close touch with this situation, obviating any unnecessary tool wastage which very quickly may become extremely expensive, and which also may cause dangerous delays in production in shop due to shortages in tool supply.

(Signed) George de Laval,
Vice-President and General Manager.

WESTINGHOUSE ELECTRIC AND MANUFACTURING CO.

East Pittsburgh, Pa.

Until recently we have confined our training, as such, very largely to trades apprentices. We have a four-year course for these as mechanics, electricians and patternmakers.

Until recently it has been our plan to train new employees in the section to which they were assigned, upon the machines and work with which they were to be regularly connected. The various shop departments have instructors picked from the best workmen to demonstrate machines and their operation, when necessary.

What has already been said applies to both men and women, and up to the present time we have done nothing more than this in the way of training men employees.

Recently we have inaugurated a training course for women machine operators and a department has been equipped wherein these women are instructed in the operation of machine tools, such as lathes, drill press, screw machines, grinders and milling machines. This department is also to train women who are to work on mechanical fitting.

In addition to this we have a training school for instructing women employees in electrical work, such as winding, taping, soldering, connecting and insulating.

These schools are primarily for beginners and it is the plan to obtain women for these training sections through our centralized Employment Department. Then, when any manufacturing department wishes help, it will obtain it from the training section.

The instructor from these training sections is a high-grade man assisted by women, and these instructors are carefully selected from our own best employees.

The length of the training period runs from two to three days up to three to four weeks, depending upon the difficulty of the occupation and the adaptability of the woman being trained.

We have also recently inaugurated a training school for women clerks and we have had for some time a training school for stenographers, typists, tracers and dictaphone operators.

We believe that preliminary training is very desirable for both men and women and that if equipment and space are allowable, special training departments should be established wherever the nature of the work will permit it.

(Signed) Robt. L. Wilson,
Assistant General Superintendent.

NORTON GRINDING COMPANY

Below is a list of men from the Norton Co., Worcester, taking a summer course in the Worcester Trade School to qualify in an eight weeks' course as all round machine hands. Five other factories are sending similar groups of employees. Their wages will soon be from 30 per cent. to 100 per cent. more than ever before, which indicates in a measure their increased service in the war emergency.

Could anything indicate better than this group in training for all round machine hands, and the statement of their previous experience which follows, the necessity of bringing men like them into our factories, first giving them an intensive training which develops their latent abilities?

Name	Age	Previous education	Previous occupation	Former wage
A. B.	24	Grammar School	Carpenter	Not stated
E. R.	32	Grammar School	Laborer	$18 per week
C. M.	30	High School, 1 year	Musician, 9 years	$22 or $20 week
C. J.	24	Grammar School	Elastic Dept.	35½c per hour and piece work
O. L.	18	Trade School, 4 years	Student	
I. K.	24	High School, 1½ years	Scraper hand	45c per hour
C. D.	20	High School, 3 years	Grocery clerk	$2.75 per day
T. J.	44	Grammar School	Painter	34c per hour
A. S.	27	High School, 4 years	Adjusting glasses in optician's office	5 yrs. at $18 week 9 mos. at $25 week
D. S.	20	High School, 1 year	Assembling, machine shop	34c per hour
M. O'B.	42	Grammar School	Brushmaker for 33 yrs.	$4 per day
C. K.	32	High School, 2 years	Assembling, machine shop	35c per hour
O. G.	19	Not given	Grading	35c per hour

S. G.	31	Norwich Univ. graduate	Civil engineer, in business for himself	
M. W.	17	High School, 1 year	Turret lathe operator, 2 years	
		Trade School, 1 year		27½c per hour
J. S.	21	Primary School	Moulder, 1 year	$3.50 per day
L. B.	20	High School, did not graduate	Grocery clerk	Not stated
M. H.	18	High School	Student	
M. T.	33	Primary School		35c per hour
A. S.	38	Grammar School	Scene shifter in Poli's Theater	$21 per week
I. G.	19	Grammar School, 2 years	Student	
H. M.	20	Not given	Carrying boxes, Logan, Swift & Brigham	30c per hour

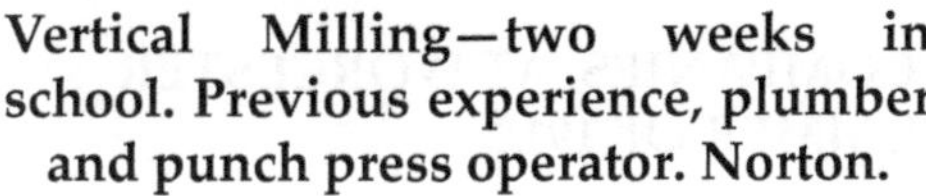

Vertical Milling—two weeks in school. Previous experience, plumber and punch press operator. Norton.

Sharpening Milling Cutter on Norton Tool Grinder—seven weeks in school. Previous experience, two years boiler factory and teamster. Norton.

MACHINISTS' CLASS, NORTON COMPANIES, AT WORCESTER TRADE SCHOOL, SUMMER OF 1918.

MACHINISTS' CLASS, NORTON COMPANIES, AT WORCESTER TRADE SCHOOL, SUMMER OF 1918.

Taking an eight-weeks' course of intensive training fitting them to take places in a general machine shop. This course carries them much farther than machine operators.

FEMALE EMPLOYEES AT NORTON COMPANY, WORCESTER, MASS., JULY, 1918, IN KHAKI UNIFORM.

Grinding on Brown & Sharpe universal machine; 5 weeks in school; previous experience, designer in corset shop (Norton).

Operating Warner & Swasey turret lathe; no previous experience in machine shop; 8 weeks in school and now earns 45 cents per hour p. w. (Norton).

Learning bench work; one of these men in school 6 weeks; previous to coming here owned a butcher shop (Norton).

DETROIT STEEL PRODUCTS CO.

Detroit, Michigan

Some time ago it became necessary to increase the number of skilled mechanics in the steel spring business and a company in Detroit segregated a group of handymen under a skilled mechanic to see if they could acquire the necessary skill in a period less than three years which was then the admitted necessary time for complete apprenticeship. The plan had the assistance of a man trained in teaching methods but without large knowledge of the special business in hand. After four weeks' training an operator was put on production work who equalled the average mechanic in this line of work and who in three months was leading the production of the shop. This initial success led to the establishment of schooling as the proper method of securing skilled help in this factory.

The elements most desirable in industrial training are:

1. Separation of the training department from production so as to avoid interference with production and also interference with the operators in training.

2. Supplying a full knowledge of the special business or trade through an instructor fully trained.

3. Supplying a full knowledge of teaching methods through an instructor who can act as a vehicle for the transfer of trade information in simple language to the new operators.

4. Application of the students' time to learning one simple operation, preferably on a subdivided operation.

5. Application of the student's time to learning the free operation of the tool together with opportunity to try its operation on some personally chosen work.

6. Satisfactory high scale of wage, but one which can be exceeded in actual production.

7. Encouragement to the operators and then more encouragement.

8. Selection of suitable operators.

Of all these the ones most overlooked are numbers 1, 3 and 6.

Many managers think that good instruction can be given right in the producing departments, whereas this has been proven to be a great interference with both production and schooling.

Most attempts to use skilled mechanics as instructors has failed because they lacked the ability to properly convey to others the knowledge they possessed. The assistance of a trained teacher has made the work of many mechanical instructors a real success.

The opinion that the company is doing a disinterested thing in training new operators has led some to believe that the wage scale for learners could be made very low. One way to help a man to act like a gentleman is to dress him as such and treat him as far as possible as such. The same holds true with operators in training. The fact that they are rated well and trained by the best of mechanics, on the best of tools, in a shop with good surroundings, means much in the final success of industrial training.

(Signed) R. S. DRUMMOND,
Formerly Vice-President and General Manager.

SCOVILL MANUFACTURING CO.

Waterbury, Conn.

The training room of the Scovill Manufacturing Company started April 1, 1918.

We train beginners on hand screw machines and engine lathes on plain turning. The training for experienced workers is to teach toolsetters with some experience to be experts along special lines and otherwise developed in their work. Also workmen with some general experience in machine room work are taught to run engine lathes. Further developments in general machine room work is to be taken up later.

Our best instructors are picked from those engaged in actual production or from promising pupils in the training school.

Skill, patience and teaching ability are the requirements of the teachers.

The best trainees are those recruited from other lines of work in the factory, especially at this time, and those impelled with the real sense of duty. Requirements: Average strength, intelligence and a desire to learn.

The steady type is preferable to the more brilliant operator who lacks staying qualities. The operators are trained in the class of work they are expected to follow, and this training is valued in proportion as it increases production from the first in the production rooms, and enables the operator to face actual working conditions without hesitancy and without fear of handling the machines.

The total cost of installation for our school to date has been approximately $2,000. With us, the cost of training (being the amount paid operatives above their earnings while in the training room) is approximately as follows:

Engine lathe workers $34 average.

Toolsetters $25 average.

Female screw machine operators $10 average.

The average number of female operators in the school is nine and their average length of time for training is eight days.

The male operators, both engine lathe workers and toolsetters, require from three to six weeks' training before they are sent out to the production rooms.

For tool room purposes we have not taken up the training of women and have only taken up the training of men along the lines of simple punch turning and straight work. We find that the men we instruct in this line of work are very interested and stick closer to the job than the average apprentice in the tool room.

(Signed) WM. COLINA.

THE RECORDING & COMPUTING MACHINES CO.

Dayton, Ohio

Several years ago we had over 200 toolmakers in our tool room engaged upon high grade jigs, fixtures, gauges, etc. The demand for toolmakers became such that the men were leaving us and it became practically impossible to get an adequate supply of this highly skilled labor.

My engineers, superintendents and myself made a study of the proposition and found that on the work that the tool room was doing it was unnecessary to employ such highly skilled labor on 70 per cent. of the work on the average. We, therefore, differentiated the work into its component elements and made a careful line of cleavage between the highly skilled work which the toolmakers were doing and the work which could be done by ordinary machinists. We then brought in men who were machinists, separating them into several necessary grades. We had sufficient work of a minor character to keep the lower grades busy practically all the time. We, therefore, taught them just how we wanted the work done.

As a result of this differentiation of the elements going to make up tool room work and the shaping of a distinct line of cleavage between the work requiring high skill and that requiring skill of a lesser grade, we were able to reduce our toolmaking force to less than fifty.

I am sure that a close study of the work done in any tool room and a division of the work same as along the lines indicated above will result in a decrease of the number of toolmakers required.

August 7, 1918.

(Signed) C. U. Carpenter.

OHMER FARE REGISTER COMPANY

Dayton, Ohio

The training department occupies a space of 25x60 feet and has the following equipment installed as a beginning:

1 13-inch lathe.
1 20-inch lathe.
1 36-inch lathe.
1 No. 5 Cincinnati Milling Machine.
1 No. 24 Osterlein Milling Machine.
1 No. 5 Brown & Sharpe Vertical Milling Machine.
1 24-inch Shaper.
1 Bathe Universal Grinder.
1 4-foot Cincinnati Bickford Radial Drill.
1 20-inch Barnes Drill.
1 Brown & Sharpe Hand Screw Machine.
1 14-inch Wet Tool Grinder.
About 30 feet of benches with vises, etc.

At one end of the space they have an office and class room, 15x20 feet. In it they have chairs, blackboard, drawing board, etc. It is their practice to assemble all of the students in the class room for a few minutes each day and give them short talks about the work and the fundamentals of the business. These talks are made as pithy as possible and only one main fact is presented at a time.

They are taking in green help, either from the laborers in the shop or hired from outside, both men and women, and are training them for machine operators and bench hands. Their conditions are such that they cannot do as many concerns do, train for a single operation, as they must make all-around operators.

Their method is such that if the foreman of the lathe department is in need of a man he makes out a "request for help" form and has it sent to the school where his needs are supplied if possible; the "request for help" is then sent to the Employment Agent stating that the request has been filled and the Employment Agent fills the vacancy in the school.

In teaching the names of parts of the various machines, they are going to give each student a picture of the particular machine he is to work on; these pictures are numbered and on a separate sheet are the names of the corresponding parts. This is done so that they can be examined in the names of parts and not have the name in front of them to refer to.

Only regular factory production which must pass inspection is used for instruction.

In regard to instructors, they have taken a man from the tool room who is a mechanic and a good teacher. He can handle any and all of the machines and has the ability to tell what he knows in a clear way that is readily understood.

Most of the students in the training room at present have been hired from outside but as it is becoming better known among the men, the laborers are applying for admission in rapidly increasing numbers. They have several traveling salesmen, office men and a few who have taken their degree. These latter are not very satisfactory, however—they are not nearly so amenable to instruction as are men who have been brought up to work.

Their training period will probably extend from four or five days to as many weeks, depending on the adaptability of the student and the difficulty of the machine for which they are being trained.

Women have not yet been introduced on the heavier machines but it is intended to do so within the near future.

COURSE FOR LATHE OPERATOR

Names of parts of machine.
Names of cutting tools and their uses.
How to set tool properly and why.
Measuring instruments, uses and how to read.
Reading blue-prints.
Tools used about and in conjunction with lathe.
Files and filing.
Starting and stopping machine.
Changing spindle speed, back gear, etc.
Starting and stopping various feeds.
Changing feeds.
Centering round shafting.
Plain turning operations.
Face plate work, how set and centered.
Grinding cutting tools, clearance, rake, etc., and reasons therefor.
Cutting speeds and feed for various metals.
Lubricants and coolants, use and benefits.
Care and upkeep of machine.
Kinks and pointers.

(Signed) E. M. Pierce,
Supervisor of School.

BURROUGHS ADDING MACHINE COMPANY

Detroit, Mich.

The Burroughs Adding Machine Company established in 1907 an apprentice school (course four years) which from its inception has proved an unqualified success. Our apprentices have also attended classes in the Cass Technical High School in Detroit. In the year 1916 we instituted a course along similar lines to our apprentice course for our service men.

However, the general shortage of skilled male help, the loss of over 800 men through the draft, and the rapid expansion of our business has obliged us to supplement our force with a considerable amount of female help in order that the increased demand for our labor-saving product be met.

Early in the present year, therefore, we established a school for unskilled female labor in connection with one of our departments engaged in the simpler operations. As the young women pass through the Employment Department they are placed in this Training School under the supervision of a competent instructor and are thoroughly grounded in the operation performed in that particular department. While in this school their characteristics are studied and as they acquire proficiency and their ability develops, they are assigned to more intricate and important work in the other departments throughout the factory. The selection for these assignments is determined by their physical condition and their mechanical development and aptitude. The instructor explains thoroughly the nature of the new employment, points out the advantages accruing to the employees because of their increased earning capacities; introduces them into the new department, points out in detail the various operations conducted therein, and painstakingly explains the scope of their new duties.

The following day they are started at their new operation, and by frequent observation, instruction and encouragements improve to a degree where they become expert in the one operation.

In this manner girls are gradually developed from the simpler burring and filing operations until we now employ them in departments performing such varied operations as indicated below.

Spring-winding, riveting machines, drill-press and milling machines, straightening of parts, assembling of special features, assembling and fitting type, the erection of machines, adjusting and inspecting machines, assembling and adjusting motors, punch press and hand and automatic screw machine work.

As the girls graduate from the starting department, or school, they don the regular shop uniform, consisting of a suit of overalls, and take their place alongside the men and under the same general conditions as to hours of labor and rates of pay. This stepping-up method of training the unskilled females has been a success with us as far as it goes, and *has enabled us to increase our production 50 per cent. for the current year in spite of the acute skilled labor situation*.

From April 1, when the training school was established, up to the present time, 412 young women have been received in Department 35, and 260 have been trained and transferred to other departments. At all times there are about forty or fifty young women undergoing training. Only nine young women have been returned to Department 35 for further training since April 1. After receiving additional training these nine were again placed and in no case has one failed for the second time. It is just a matter of finding the right place for the right young women, and then there is no question about them making good on the jobs, as they are proving every day.

In conclusion, tribute must be paid to the 1,200 women in our factory whose earnest desire to help their country in its time of need, and whose mentality and courage have enabled them to make a success of a kind of employment entirely foreign to them on the general conception of their abilities.

(Signed) WM. EARL LEEVER,
Assistant to General Manager.

UNDERWOOD TYPEWRITER CO., INC.

Hartford, Conn.

The Underwood Typewriter Co., Hartford, Conn., has undertaken the employment of women on a part-time basis, such as will permit them to attend to their household cares to a reasonable extent. Further, they are offering employment to women having small children between two and one-half and nine years of age, having given over a space in their plant for the care of such children throughout the work day, practicing the kindergarten plan. They have found many who are willing to engage with them under this plan, and are pleased to report the whole general scheme is working out well. Many of the women of either class have become expert in skilled work with but a limited time for training. Under their method, however, the instructing is done in each of the manufacturing departments where the plan has been introduced, as they have operated under good regulations as to quality and quantity for many years back, rendering it very practicable in their case to not instruct and train in separate spaces, although they appreciate the need for acting otherwise with new work, such as has been brought about by the war, and wherein the tasks at hand are not subject to accurate measurement to start with.

(Signed) C. D. RICE,
Manager of Factory.

Photos are herewith shown which illustrate the mothers on part-time work in the factory while the children attend the kindergarten in the plant under expert instruction.

The experience of the Underwood Company indicates conclusively that advanced age is no barrier to productive value.

Part-time workers assembling typewriters. Their children are cared for meantime in the company's kindergarten—Underwood Typewriter Co.

War-time kindergarten. Underwood Typewriter Co.

THE NATIONAL CASH REGISTER COMPANY

Dayton, Ohio

The Training School for women, of whom over eight hundred have been placed in the various departments, was started in the latter part of March, 1918.

Because of the demand for trained help in the factory, we have not been able to keep them in the Training School as long as we would wish, but even this short experience has been sufficient to take away the fear of the shop, as many of our women have never had any factory experience before.

While in the Training School the students are paid the regular starting rate for women, and after they enter the factory and become more efficient their rate increases until they can do the work that a man previously did both as regards quality and quantity and they receive a man's wage.

In some departments it has been found necessary to put slightly more women on the same operations than men formerly employed to obtain the same production, but as the women gain experience, their production increases rapidly and the quality is as good, if not better. The women have proven themselves very apt in picking up the smaller class of assembling on account of their nimble fingers and care in handling stock.

We use our regular production to train the students, and as it must pass 100 per cent. inspection, we emphasize quality and not quantity.

We find the best class of workers comes from those between twenty-one and forty years of age, with, of course, exceptions.

We not only teach the new employees the mechanical operations, but also give them "Health and Safety" lectures, and show them pictures of many ways one can become injured if they do not use precaution while working around machinery. They are also instructed in the use of time and instruction tickets.

Our Inspectors are selected from the factory, preference being given to those who are experts on their particular class of work.

It is our opinion that the Training School is the proper way to teach the inexperienced help in order that they may learn the work quickly and get on a production basis in a short time instead of hiring and placing help right in the shop and letting them pick it up with what assistance and instruction they can from their fellow workmen.

In the Training School the most efficient workers can be reorganized in a short time, and the less efficient ones can be given special attention, and usually *we can bring them to a degree of efficiency not possible under the old method.*

We try to find out in the training school where the student's strong point is, whether on machine operation or bench work, and are enabled in this way to place them in a job they are particularly suited for, thus keeping the problem out of the factory. If they show a proper degree of interest, they are given all possible encouragement.

We have had to materially increase the size of our school, and with the hearty co-operation we are receiving from the heads of the different departments, we believe the employment of women on our work is proving a success in every way.

Another point that we think is good is that the school itself is nearly self-sustaining.

The accompanying photo, number one, will give an idea of the size of our school, and also the various classes of work we train them on.

Number two shows a gang of hand mills "manned" by women. These girls have all been through our Training School and are now working on a piece basis and doing it successfully.

(Signed) WM. A. HARTMAN.

No. 1. National Cash Register Co. A section of the training room.

No. 2. National Cash Register Co. A gang of hand mills operated by women.

PACKARD MOTOR CAR COMPANY

Detroit, Mich.

In the spring of 1914 labor conditions were somewhat disturbed in Detroit. We lost a good many of our expert varnish rubbers, and we could not get skilled men to replace them, and we tried to break in men on the varnish rubbing deck, but found that too much work was spoiled by the green men, and the experienced men did not have time or inclination to properly instruct those who were unskilled. This led to the establishment of a school for training varnish rubbers and was the beginning of our efforts to train unskilled workers. The result of this experience was so highly successful that we carried it to all of the other branches of body manufacture, and *a school for training unskilled help became a permanent part of our institution.*

We were able to teach women how to trim automobile bodies, and they learned in an average of less than ninety days. Their work was of a very high order, and we were very much gratified with what they accomplished. Very often we found that we were able to train men to an exceedingly high degree of skill in less than sixty days. All men were not so apt. Some of them could not be placed in the skilled class even after three months of training.

As the war activities took workers we turned our school into training quarters for mechanics. For instructors we aim to use the best workers in a particular class that we have, providing that they have the natural ability to instruct others. Some men lack the ability to impart their knowledge to others. This type of man does not make a good instructor. A man should not only be an expert in his trade, but he should have the natural ability to impart his knowledge in order to become a good instructor.

In teaching some of the women we find that some women pick up very quickly some particular trade and when they become experts we in turn make them instructors. We have not yet developed any toolmakers. We have developed men to tool room work, such as turning, grinding, etc., but this is not really toolmaking.

We believe that it is possible to turn out good toolmakers and we are turning our energies to this end.

(Signed) F. F. BEALL,
Vice-President of Manufacturing.

NORTH AMERICAN MOTORS CO.

Pottstown, Pa.

Our production is about half of what it should be, due to lack of skilled help, in other words, due to lack of machinists. We have not had much trouble in getting green men and we have had no trouble in training these green men to do the operations on the shell both accurately and quickly, but we have had trouble in *getting men to maintain the tools or to equip for these operators.*

Facing this condition and being unable to obtain machinists, we decided to train men to do machinists' and toolmakers' work.

Our scheme as outlined is as follows: We are taking operators who have had experience of a year or more in our shop and are putting them *into the school* under a good mechanic who fortunately is a teacher also. After a few weeks in the school we are *putting them into the tool room as operators,* that is, they will be on work where they will get work *of a repeating nature* and they will stay on the same class of work for a considerable period of time, depending on the man and how fast our other pupils come on, the idea being then to take the first man back again to the school and teach him to operate some other machine tool, then send him back in the tool room, where, after he has operated on the second type of tool for a certain length of time, he would then be of more use, as the tool room foreman could then place him on either one of two machines. For the more attentive and interested men we would continue this scheme and thus teach them the operations of all machines and tools used in the tool room. We have also provided for a certain line of bench training.

Within the next few weeks we hope to be able to take some of our machine tools from the shell shop and put them into this school and we will then train our operators in this school with the idea in mind, as stated in our previous letter, of teaching them the proper care of a machine tool, as we consider this of vital importance. In fact, the writer would say that from observation in other plants and experience here that it is his opinion that one of the greatest, if not the greatest, causes for lack of production in machine shops now on shell work is due to *machine breakage,* this coming from several causes, among the foremost being the lack of knowledge on the operators' part.

When the idea of the school was first brought up there was some feeling among the skilled mechanics that the men trained in the school would replace them to the detriment of the mechanic, but this idea has passed or is passing away very rapidly and we find a considerable interest shown by the mechanics in the things that we are teaching. Two of our good machinists who are on maintenance or repair work have asked to be allowed to go to the school so that they may become better workers and get a training on finer work. Others have shown similar interest and we have tentatively agreed to start an evening school this fall for mechanics, our proposal being to work in conjunction with the Y. M. C. A. for shop drawing and the reading of drawings and to use our own shop for special instruction on machine tools.

(Signed) George C. Lees,
Secretary and Works Manager.

THE H. E. HARRIS ENGINEERING CO.

Bridgeport, Conn.

We are enclosing herewith three photographs showing work in our vestibule school for women on gauge finishing work. The picture of the six women in a line with the instructor at the end shows the pupils on gauge finishing work, lapping plug, thread gauges, snap gauges, etc. These women have proven themselves very apt, but difficulty is being experienced, due to the fact that the necessary laps requiring highly skilled mechanics are often made defective, on account of the feeling of the skilled tool and gauge makers who do not wish any of the women to do any of this work.

The women show a better spirit and give a much better production, at least three times as much as the men do on the same work. The one photograph showing the six women in a row and the instructor at the far end, shows a group in the school lapping these gauges.

Group in training room lapping gauges. H. E. Harris Engineering Co.

The two photographs of the same woman, Mrs. H— —, show her in one photograph lapping a thread gauge which has to be correct within .0002. She is about four times as proficient as any man that we have in the place. The other photograph shows her measuring the same thread gauge between the lapping operation with the three-wire system, which is rather a difficult feat of measurement.

(Signed) HARRY E. HARRIS,
President.

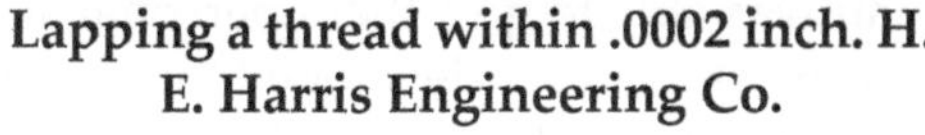

Lapping a thread within .0002 inch. H. E. Harris Engineering Co.

Measuring thread gauge with three wire system—a difficult feat. H. E. Harris Engineering Co.

OAKLEY MACHINE TOOL COMPANY

Cincinnati, Ohio

Our shop has an average of 75 men on its pay roll, making a Precision Tool Room Grinder. Before the United States joined the forces of Liberty we had felt a shortage of labor and had taken steps to break in untrained men.

Thinking we were not large enough to inaugurate a Vestibule Training Room, as it is generally understood, we inaugurated a system of training men directly in our shop.

We sorted out applicants and put them in our shop, two at a time; if they had never worked in a shop before we started them on simple machines, such as the hack saw, centering machine, etc., in order that they might get used to the noise and methods of the machine shop. They were then advanced to Roughing Lathes, being given simple jobs, such as turning and facing.

By having only two at a time the foreman was able to give them personal supervision, without interfering with his regular work. As they developed they were given more difficult jobs. We found, as a rule, inside of sixty days such men made very fair machine hands.

We also broke men in on drill presses and shapers, using same tactics and had very successful results. To give you an idea as to the class of men from which we have made machine operators, we have working in our shop to-day one bartender, piano tuner, street car conductor, bricklayer, coal miner and an artist—self made. The other unskilled men had had some previous experience on productive labor, either running punching presses, nailing machines, or work requiring a smattering of mechanical ability.

Our experience is that if you take a man over 30 that has become disgusted from a blind alley profession, where there is no hope of advancement, point out the possibilities of the machine tool trade, and give him a living wage to start, even though at first he is not worth it, he develops into a good and loyal man. They are, however, like children, they have to be encouraged every so often by a personal talk or suggestions from the head man.

Of course we have had our failures, but our successes have been in the majority, so we are continuing to break in green help.

(Signed) Albert A. Thayer,
Treasurer.

LINCOLN MOTOR COMPANY

Detroit, Mich.

Our school is going along nicely and while we are not perfecting machine tool operators to the degree I would like because of the necessity of rushing them through the school to the shop proper, we are accomplishing, I think, that which we set out to do, namely, to take away from the girl the fear of the shop and to give her a fair knowledge of the tool she is to handle. The women undoubtedly have benefited beyond measure by the short time spent in the school room, and have gone into the factory with the confidence that carried them through the first few days and made them efficiently productive in a shorter period of time.

The training room is located in the smaller of our two plants and is equipped with a lathe, milling machine, gear cutter, drill press, profiler, etc., those being the tools upon which it was decided to train operators. In charge of this room was placed an instructor who had had some slight experience in a continuation school and *who went to work under the direct supervision of a high grade specialist secured from a well known eastern factory efficiency organization.* The instructor was given no special instruction beyond being told what we hoped to accomplish in the way of building up an organization of women of more than ordinary ability and moral character.

The training room up to the present time has been used only in connection with supplying the factory with women workers. Women of the age of twenty-one and upwards have been taken, their references carefully examined, and they have been given *from one to three days' training in this school.* Because of the demand of the shop for help it has not always been possible to keep them in a training room for as long a period as would seem desirable, and in some instances they have stayed only one day.

During the training period they have been paid the regular rate for women, thirty cents per hour, which rate maintains after they enter the shop until such time as they are placed upon a piece-work basis.

We believe, however, that through the medium of the training room we shall be able to instruct women workers in machine tool operation so they will go direct from the school room into the shop without fear of what is to be encountered therein, and with a better knowledge of the tool they are operating, and the reason they are operating it, than they could possibly acquire through any other method.

Inspecting pistons and valves. Lincoln Motor Co.

Machining Main Bearing Bolts. Lincoln Motor Co.

Hand Milling Machines. Lincoln Motor Co.

In the training department. Lincoln Motor Co.

New arrivals in Training Room—Lincoln Motor Co.

New arrivals in Training Room—Lincoln Motor Co.

The school is favorably looked upon by all of the employees, and in those cases where it is found that a woman is not working out well upon the work to which the school has assigned her, and is returned to it for further instruction, she has in all cases gone back to it with a cheerfulness and willingness that is both surprising and gratifying.

The writer is of the opinion that the school in this factory has come to stay and that when we build up our organization and get through the strenuous times we are now experiencing, the advantage of the vestibule school instruction will be given men employees as well as women.

(Signed) J. M. Eaton,
Assistant to President Henry M. Leland.

TRAINING FOR THE TOOL ROOM

You mention training in the more difficult lines of effort, such as Tool Room, Machine Shops, etc. We agree with you as to the wide possibilities in this field, and are now organizing to accomplish this very thing at the Lincoln Motor Company.

At the beginning of the war when labor shortage was seriously manifested, I had charge of the reconstruction of a tool room, employing 270 toolmakers and machinists, engaged in the making of cutters, reamers, broaches, drill jigs, milling fixtures, etc.

Realizing then the difficulty that the future held in store of securing competent, reliable toolmakers and machinists, we undertook to train men who had no previous experience in this line. Our results were quite gratifying. We classified all work and trained men to operate specific machines against the various classifications.

To illustrate: We engaged a carpenter 67 years of age, who had no previous machine shop experience, trained him to run a Universal Milling Machine; not only did he meet successfully all work scheduled against his machine, but developed such skill in the operation of this machine, that later he compared favorably with the average good toolmaker, and in six months' time we depended on him entirely to set up his own machine and proceed as a regular toolmaker. This represents intensive training and what was accomplished here can be accomplished in other general lines of tool room work.

We ultimately had 215 of these trained operators who were able to carry on the work, supported by 55 good toolmakers. I am with you in the confidence that it will be necessary to plan quite considerably in the adoption of some means to overcome a situation which looks really serious for the future.

(Signed) W. H. Ebelhare,
General Superintendent, Lincoln Motor Co.

MUELLER METALS COMPANY

Port Huron, Mich.

I am pleased to advise you that we have been using many unskilled workers, also women, at both our Sarnia, Ontario, and Port Huron, Michigan, and Decatur, Illinois, plants and find that women are able to do the light operations on turret lathes quite satisfactorily.

Our greatest difficulty is in our Toolmaking Department, but we have lately installed the following plan.

We are selecting a good lathe hand from among our toolmakers and giving him from one to three students, paying him from 20 per cent. to 30 per cent., depending on the number of students he is able to take care of successfully, and we then pay these students about 50 per cent. of what the instructor receives with stipulated raises in pay until they have served two years, at the end of which time we pay them a very liberal bonus.

We have been able to secure some young men just out of High School who are going to make good workmen, but as the new draft will possibly take some of these boys we are now figuring on using women on this work and believe that they will be able to carry it on quite successfully.

(Signed) C. G. Heiby,
Vice-President and General Superintendent.

DETROIT LUBRICATOR COMPANY

Detroit, Michigan

New employees are started on simple work and given individual instruction. They are then advanced to more difficult work.

By this method of training we have operators giving top production who only a short time ago were automobile salesmen, cigar salesmen, letter carriers and watchmen.

The women trained in this way are producing excellent results and are making as good pay as the men on the same piece-work. At some types of inspection they excel any men we ever had on the jobs for speed and accuracy.

We have two men, with *only one arm* each! (after proper training) doing more work than the average two men.

The company has created a new department for the production of Liberty Motor Aeroplane Carburetors, and the work of this department includes a great variety of machines. It writes:

In order to furnish employees for this work with at least partial training, we are recruiting them from other departments in the shop doing similar work; in each of these other departments we have a section set aside which includes type machinery, and new employees are put at this work under careful supervision and training to develop them and thoroughly determine their value before being advanced on to aeroplane work.

(Signed) G. B. Duffield,
Superintendent.

INDEPENDENT PNEUMATIC TOOL COMPANY

Aurora, Illinois

Since installing our Vestibule Training School some seven weeks ago, we can gladly say it has proven a success, even beyond our expectations.

To date we have enrolled 109 students; 81 men, 28 women. The women workers have proven that they can in the emergency take the place of practically all our male workers, that is with from five to ten days of intensive training. We have placed women on such machines as Gear Hobbers, Screw Machines, Grinders, Drill Presses and are well satisfied with the result obtained. The percentage of scrap material has been less by the female workers than by the male.

In beginning our course of training each week our Mr. F. B. Hamerly, Works Manager, has made it his duty to tell these workers the vital need of training unskilled men and women so as to replace the boys that are called from our plant, that is, he tries to raise them to the highest pitch of enthusiasm.

(Signed) A. H. Boehm,
Vestibule Training School.

August 16, 1918.

INDEPENDENT PNEUMATIC TOOL COMPANY.

In training on regular production.

Our training school has been running about five weeks with a capacity of 12 students per week. We are indeed surprised at the results we are getting.

Next week we intend to put on a double dose running through 24 students per week. We are trying to start a night shift in one of our important departments and will need about 48 or 50 men. These men are to be recruited entirely from our Vestibule School, where in the past we have been playing in the open market competing for the desired help.

We are satisfied that if the larger corporations and manufacturers were to install a Vestibule system in their factory that within the next six months the severe competition for skilled mechanics would be practically eliminated, and production would be increased considerably.

(Signed) F. B. Harnelly,
Works Manager.

ROYAL TYPEWRITER CO.

Hartford, Conn.

REPORT OF TRAINING SCHOOL FOR JULY, 1918

Total entries since July 1	100
Total number of operations taught	18
Total number of permanent instructors	6
Instructors from other departments	5
Total number transferred to other departments	73
Total number left factory from school	13
Total number in school August 1, 1918	14

NAMES OF OPERATIONS TAUGHT IN TRAINING SCHOOL

Milling	Light Power Press
Heavy Power Press	Small Parts to Base Assembly
Copper Bar Grinding	Shift Arms and Bottom Rails to Base Fitting
Brackets to Base Assembly	Keycard Assembly
Linking	Second Carriage Building Assembly
First Carriage Building Assembly	Riveting (machine)
Inspection of Base Assembly	Retapping
Drilling	Type Soldering

GENERAL

All students but three are women, ranging from fifteen to fifty-seven years of age.

Intensified training, speed and accuracy, physical and mental condition of employees, proper mental attitude to company.

A matron has been appointed who comes in contact with the women and is responsible for their deportment and personal needs.

(Signed) CHAS. B. COOK,
Vice-President.

CINCINNATI MILLING MACHINE CO.

Cincinnati, Ohio

We put the new employee into a department and assign him or her to a machine in charge of a skilled operator. The new employee becomes at once an observer and a helper, and in a little while takes charge of the machine and the skilled operator stands by and gives special instructions.

We have detail instruction cards or process sheets for all operations, and one of these cards in the hands of the new operator will serve as a guide for turning out the work properly after the instructor leaves the new employee to himself or herself.

In addition to this we have certain selected engineers from the Time Study Department, who are attached to both the day and night shifts, and devote their entire time to coaching the new operators, and see to it that they learn to acquire the desired degree of skill and proficiency.

We find that the women who are selected for this sort of work just about equal the men. They show considerable enthusiasm for the work, as is indicated by a less degree of lateness and absenteeism than that of the men, but we have not had enough experience as yet to say anything definite in this regard.

It is also perhaps true that we are taking greater pains instructing the women than we would in the ordinary course take in instructing green men.

(Signed) Charles S. Gingrich.

THE STANDARD PARTS CO.

Cleveland, Ohio

Each individual department at the start of the year had its own training division, that is, while we had the orders we lacked a good many of the machines necessary to do our work and the proper tool equipment to start with. We also required the services of eight hundred additional people.

We trained absolutely unskilled men and women during this slack period, so that when we started in quantity production we had also obtained speed. An item that might be of interest to you is the fact that we are now employing women in Drill Presses, Milling Machines, Hand and Automatic Screw Machines, Turret Lathes, Speed Lathes, Engine Lathes, Assembly Work and Inspection Work.

Automatic Milling Machine—Standard Parts Company.

These women and the majority of the unskilled men whom we employed are doing work in most cases where dimensions are held to one-half thousandth of an inch limit variation. In a plant employing excess of five thousand people am absolutely convinced that a separate vestibule training school is a necessity, and in plants already producing work in large quantities there is liable to be a heavy demand for trained skilled workers, a separate training school would be necessary.

However, where the number of workers needed do not exceed ten people on an individual operation training on machines in the department would be sufficient.

August 7, 1918.

(Signed) J. A. Rothenberg,
Employment Manager.

A skilled worker on Automatic Turret Lathe—Standard Parts Company.

THE YALE & TOWNE MFG. CO.

Stamford, Conn.

We have had in operation for over a year a vestibule school for the training of women employees in our plant, and are obtaining good results from it.

We are training the women mostly for bench and machine work which was formerly done by men, such as: Lock assembling; drill press work, which was formerly considered inappropriate for women employees; hand screw machine and automatic screw machine operators.

We are also training female help on lathe and shaper work. We plan to do the same thing on milling machines and expect eventually to include tool room work.

Our vestibule school activities include the training of male foundry workers, and the training both of men and women to become instructors and machine adjusters.

We also have an Apprenticeship School.

August 13, 1918. (Signed) The Yale & Towne Mfg. Co.,
J. A. Horner, *Vice-President.*

ILLINOIS TOOL WORKS

154 E. Erie Street, Chicago

We are planning to have this school in operation within the next month. The writer expects to be responsible for the results. We expect to take one good man from our own plant as an instructor, and we are also in communication with an instructor from a college in a Western State whom we may have in direct charge of this work should he prove to be the proper person.

At the present time we have in our factory about 75 female employees, on Lathes, Milling Machines, Grinders, Finishing Gauges, Lapping, etc., also inspectors, timekeepers and stock chasers.

Since your last visit we have employed a trained nurse who is in charge of the employment and welfare work of all women employed in the factory. This we have found has given us much better results and can truthfully say that with very few exceptions, every girl employed is certainly making good.

We have one instance where a man employed in the screw machine department, employed in that capacity for about a year, was having trouble in not producing on his machine. We had him exchange machines with a woman who had had a month's experience and found that she practically doubled his output the first day.

(Signed) J. D. Sherman, *Factory Manager*.

August 5, 1918.

GLEASON WORKS

Rochester, N. Y.

When you came to Rochester we were very much impressed with the suggestions you made as to the introduction of women into industry. We sent two representatives together with others from Rochester to Dayton, as you advised, to investigate the conditions there and also in Cincinnati. The excellent arrangements made by the Cincinnati manufacturers to relieve the shortage of labor by placing women at work in machine shops and elsewhere were extremely interesting.

After learning what had been done we started using women in our machine shop in line with your idea and the results have been very satisfactory. * * * It is not a question of economy with us but of releasing men for other work in the foundry which women cannot perform. We believe that an intensive training of two weeks would enable women to turn out practically as much work as men are now doing.

August 5, 1918.

(Signed) JAMES E. GLEASON.

DIAMOND CHAIN AND MANUFACTURING COMPANY

Indianapolis, Indiana

I am sure that you will be interested in knowing that we have already undertaken this work. For some weeks we have had a school going on for the training of screw machine operators with every success. We also have here a man who is giving detailed attention to the personnel in connection with our hardening and cyanide work. He is talking to these men individually and in small groups, thus instilling into them the fundamental principles of heat treatment, which is making a marked impress on the character of our product.

We are also having a detailed study of the operations now being done by women with the idea in mind to substitute women where possible when it becomes necessary and we have here two women studying our operations preparatory to giving instructions to women as to how to perform the operations that men are now doing.

July 25, 1918. (Signed) L. W. WALLACE.

HENRY DISSTON & SONS

Philadelphia, Pa.

We have established a school for vocational training, appointing an instructor in each department who takes direct charge of each new employee. Through this system we have placed many women on work which was entirely operated by men heretofore.

One of the delicate situations confronting us at this time is the transferring of help from one department to another. We are appealing to their patriotism and have established a Disston Volunteer Transfer System—Industrial Soldiers Volunteering for Heavier Work, Filling Jobs Where Women Cannot. We are very much interested in this work and are making every endeavor to eliminate the labor turnover which we are all experiencing in these times. We have women operating milling machines, drill presses, emery grinding machines, file hardening, saw setting and filing machines, also power punching presses, all work never attempted by women in our factory before. The result has been more than satisfactory.

August 7, 1918. (Signed) Wm. D. Disston,
Vice-President.

THE GRATON & KNIGHT MANUFACTURING COMPANY

Worcester, Mass.

We do not maintain a regular separate school for training the employees on our business because of the fact that we have such a large variety of trades that it would not be practical for us. We do, however, maintain a training system for the employees in our various departments and have men assigned for such work in the departments.

When we have a large number of employees to train in some one line of work we establish a separate organization for them with a view to training them to carry on very efficiently and become skilled employees as soon as possible.

August 7, 1918. (Signed) F. H. Willard, *Assistant General Manager.*

THE SPARKS-WITHINGTON COMPANY

Jackson, Mich.

We are very glad to be able to state that we are operating a school especially for women to train them for toolmaking.

We have found that there is not a sufficient number of experienced toolmakers to meet the demand and the only way is to have the women help out, and we have a school for this purpose, also have been teaching them to operate production machines.

July 31, 1918.

(Signed) W. J. Corbett,
Assistant Manager.

LOCOMOBILE COMPANY OF AMERICA

Bridgeport, Conn.

The writer was in Detroit a few weeks ago and was much interested in the work being done by Packard and Lincoln with women that were trained in their schools.

We are confident that much good will result from the co-operative plan that you have adopted in developing operators. We wish to be considered among those who approve of this plan and we will arrange for a Vestibule School or Training Room as soon as we can do so. We have been training young men and boys from our local high school as junior toolmakers, with excellent results.

July 10, 1918.

(Signed) H. H. Edge,
Factory Manager.

THE TIMKEN ROLLER BEARING CO.

Canton, Ohio

We have established a Training School which has been in operation for the past two or three weeks.

We do not feel as yet that we want to make any extensive comments upon the results secured, but we believe the idea is fundamentally correct, and hope to have recorded sufficient data of interest in regard to the school so that we will be in a position to answer any questions that may be asked us.

(Signed) The Timken Roller Bearing Co.,
F. T. Mackay, *Employment Manager.*

July 18, 1918.

CARLTON MACHINE TOOL COMPANY

Cincinnati, Ohio

For your information we wish to advise that we have started a training school for girls in our plant, and we find that it is working very satisfactorily and we expect to have in the future 50 to 75 per cent. of our work completed by girls. The writer has noted your talks and writings on this subject and is trying to follow out your ideas as near as possible.

August 9, 1918. (Signed) JACK C. CARLTON.

THE WILLYS-MORROW COMPANY, INC.

Elmira, N. Y.

We have about twenty women learning how to run automatic screwing machines. These women have been grinding their own tools.

August 7, 1918. (Signed) J. E. MORROW,
Secretary and General Manager of Production.

CROMPTON & KNOWLES LOOM WORKS

Worcester, Mass.

Our training school is working very satisfactorily, and we are only sorry that we did not start two years sooner.

July 13, 1918.

(Signed) H. L. Robinson,
Employment Service Department.

THE LODGE & SHIPLEY MACHINE TOOL CO.

Cincinnati, Ohio

In the Tool Room we are using men that are skilled *in one class* of work on *one type* of machine for the majority of our tool room work.

August 2, 1918. (Signed) Joseph T. Wright,
Assistant Works Manager.

THE STENOTYPE COMPANY

Indianapolis, Ind.

We expect very shortly to have our training school in operation. We are thoroughly convinced of the advisability of establishing such a school.

August 16, 1918.

(Signed) R. M. Bowen,
Chairman, Board of Directors.

INTERESTING PARAGRAPHS

Says Mr. E. G. Allen, the able Director of the Cass Technical Trade School, Detroit: "We have taken high grade machinists in Detroit who have been in the shops for a couple of years and were familiar with the use of drawings, decimal equivalents, etc., and made tool room machine operators doing work of considerable variety, each on a single type machine, almost immediately. In three or four months, by continuing to watch and instruct such a man, he has been able to run almost any machine and do on it almost any work laid out by the toolmaker."

* * * * *

This war is going to last years. Even three or four months, which seems a long time, will pass like a day. Are some of us not almost grossly careless in not getting at this immediately? Mr. J. J. Pierson, Dilution Officer of the British Ministry of Munitions in the London District, says: "You can make a toolroom operator of a woman in three weeks. If you can't do it in three weeks, you can't do it at all. You have simply gotten the wrong woman. Pick out a long fingered, sensitive, intelligent woman from the shop force who has been carefully trained and is especially satisfactory and exact in her production and upgrade her in this way."

* * * * *

"This office has made an exhaustive study of the vestibule training methods and results of the Section on Industrial Training, of the Council of National Defense, and believes that this general immediate adoption is absolutely essential to meet the increased war program and cannot be too quickly or extensively adopted and should have the immediate and fullest support of all who are charged with production matters. The shortage of skilled labor to-day is alone two hundred and fifty thousand and we are advised will be one million by January 1. This office is putting it into force and effect just as promptly and as actively as we know how."

August 8, 1918. (Signed) John C. Jones,

Chief of Ordnance, Philadelphia District, Philadelphia, Pa.

* * * * *

"I am confident that any community by establishing a proper school connected with the industries may become a great factor in the progress of the community's industries.

* * * * *

"We find your bulletins very useful. You are sending out a world of information which education ought to receive. This war ought to show us the way in our schools and give us a chance to connect up the educational wagon with life."

August 10, 1918. (Signed) Augustus O. Thomas,
State Superintendent of Schools, Augusta, Maine.

* * * * *

"We are making plans to introduce training schools into all Ordnance Manufacturers' plants in this district. We will endeavor to make sure that representatives of all the Ordnance plants in this neighborhood hear you and work out from the enthusiasm which no doubt your exposition will create."

Aug. 10, 1918. (Signed) B. A. Franklin, *Major, Ord. R. C.*
Bridgeport District, Bridgeport, Conn.

* * * * *

The following firms in Worcester, Mass., are using the vestibule principle in special efforts at training unskilled men in their shops: The Heald Machine Company, Bradley Car Works, Reed Prentice Company, Sleeper & Hartley, Inc., Rice, Barton & Fales.

* * * * *

In a large factory making power machines the men from one department threatened to strike because "the women were being paid higher wages than the men." Investigation disclosed that all were working at the same piece rates but the women were producing more.

* * * * *

A member of a British Commission which visited the United States last winter said:

"England delayed the winning of the war two years by delaying the introduction of women one year."

PARTIAL LIST OF VESTIBULE SCHOOLS OR TRAINING ROOMS IN FACTORIES

Recording and Computing Machines Co., C. U. Carpenter, C. P., Dayton, Ohio.

Curtiss Aeroplane Corporation, F. L. Glynn, Director of Training, Buffalo, N. Y.

Wright-Martin Aircraft Corporation, J. F. Johnson, Director of Training, New Brunswick, N. J., and Long Island City, N. Y.

Nordyke & Marmon Co., Indianapolis, Ind.

Packard Motor Car Company, Detroit, Mich.

Lincoln Motor Company, Detroit, Mich.

Norton Grinding Company, John C. Spence, Superintendent, Worcester, Mass.

NOTE.—The above training rooms are remarkably efficient and successful; they should be seen. By special arrangement the men above named are assisting in the development of training rooms in their vicinity. Write them when ready to act. Also O. D. Evans, very expert in training, Army Ordnance Department, Production Division, 1710 Market St., Philadelphia.

Gillette Safety Razor Company, Boston, Mass.

Royal Typewriter Company, Hartford, Conn.

Scovill Manufacturing Company, Waterbury, Conn.

Remington Arms Company, Bridgeport, Conn.

Bullard Engineering Company, Bridgeport, Conn.

The H. E. Harris Engineering Company, Bridgeport, Conn.

Trego Motor Company, New Haven, Conn.

Winchester Repeating Arms Company, New Haven, Conn.

Taft-Pierce Mfg. Co., Woonsocket, R. I.

Brown & Sharpe Company, Providence, R. I.

Crompton & Knowles Loom Works, Worcester, Mass.

American Steel & Wire Company, Worcester, Mass.

Norton Company, Worcester, Mass.

John Bath & Company, Worcester, Mass.

Graton & Knight Company, Worcester, Mass.

Blanchard Machine Company, W. W. Blackman, Superintendent, Cambridge, Mass.

E. W. Bliss Company (torpedo factory), Brooklyn, N. Y.

Ford Instrument Company, New York City.

Pierce-Arrow Company, Buffalo, N. Y.

King Sewing Machine Company, Buffalo, N. Y.

New York Airbrake Company, Watertown, N. Y.

Seneca Falls Mfg. Company, Seneca Falls, N. Y.

Savage Arms Company, Utica, N. Y.

Henry Disston & Sons, Philadelphia, Pa.

Fayette R. Plumb, Inc., Philadelphia, Pa.

Hess Bright Manufacturing Company, Philadelphia, Pa.

Roberts Filter Manufacturing Company, Philadelphia, Pa.

Remington Arms Company, Eddystone Plant, Philadelphia, Pa.

Naval Aircraft Factory, League Island, Philadelphia, Pa.

American International Shipbuilding Company, Hog Island, Philadelphia, Pa.

Newton Machine Tool Company, Philadelphia, Pa.

Warren Webster Company, Philadelphia, Pa.

L. H. Gilmer Company, Philadelphia, Pa.

Burke Electric Company, Erie, Pa.

Bethlehem Steel Company, Bethlehem, Pa.

J. G. Brill Company, Philadelphia, Pa.

Lanston Monotype Machine Company, Philadelphia, Pa.

Leeds & Northrup Company, Philadelphia, Pa.

David Lupton Sons Company, Philadelphia, Pa.

North American Motors Company, Pottstown, Pa.

Standard Aircraft Corporation, Elizabeth, N. J.

Snead & Company Iron Works, Jersey City, N. J.

Spicer Mfg. Company, Plainfield, N. J.

International Motors Company, Plainfield, N. J.

Woodbury Bag and Loading Company, Woodbury, N. J.

American Shell Company, Paterson, N. J.

Worthington Pump Company, Harrison, N. J.

Neptune Meter Company, Hoboken, N. J.

International Arms and Fuse Company, Bloomfield, N. J.

Thomas A. Edison Corp., Orange, N. J.

General Electric Company, Newark, N. J.

Crocker-Wheeler Company, Ampere, N. J.

Submarine Boat Corp., Newark, N. J.

Barbour Flax Spinning Company, Paterson, N. J.

American Tool Works Company, Cincinnati, Ohio.

Lodge & Shipley Machine Tool Company, Cincinnati, Ohio.

Cincinnati Milling Machine Company, Cincinnati, Ohio.

Cincinnati Grinder Company, Cincinnati, Ohio.

Cincinnati Planer Company, Cincinnati, Ohio.

Cincinnati Bickford Tool Company, Cincinnati, Ohio.

Oakley Machine Tool Company, Cincinnati, Ohio.

Buckeye Twist Drill Company, Alliance, Ohio.

Morgan Engineering Company, Alliance, Ohio.

National Cash Register Company, Dayton, Ohio.

Ohmer Fare Register Company, Dayton, Ohio.

The Timken Roller Bearing Company, Canton, Ohio.

Joseph & Feiss Company, May Thomsen, Employment Department, Cleveland, O.

Mosler Safe Company, Hamilton, Ohio.

Ford Motor Company, Detroit.

Long Manufacturing Company, Detroit.

Studebaker Corporation, Detroit.

Solvay Process Company, Detroit.

Morgan & Wright Company, Detroit.

Detroit Steel Products Company, Detroit.

Burroughs Adding Machine Company, Detroit.

Dodge Brothers, Detroit.

Detroit Lubricator Company, Detroit.

Timken Detroit Axle Company, Detroit.

Haskellite Mfg. Company, Grand Rapids, Mich.

Republic Motor Truck Company, Alma, Mich.

Sparks-Withington Company, Jackson, Mich.

Illinois Tool Works, Chicago, Ill.

Union Special Machines Company, Chicago, Ill.

Western Cartridge Company, East Alton, Ill.

Independent Pneumatic Tool Company, Aurora, Ill.

Diamond Chain & Manufacturing Company, L. W. Wallace, Indianapolis, Ind.

General Electric Company, E. H. Barnes, Superintendent, Fort Wayne, Ind.

Pawling & Harnischfeger Company, Milwaukee, Wis.

EXTRACT PROM A REPORT BY MR. BEN H. MORGAN, EXPERT ADVISER TO THE DILUTION SECTION OF THE BRITISH MINISTRY OF MUNITIONS

When war broke out in August, 1914, the Government of the day urged employers to induce their skilled as well as unskilled men to join the colors. No good purpose will be served in my characterizing in suitable terms the unwisdom of such a step. Sufficient to say that it was a long time later before the Government realized that this was an Engineers' war. The result was that the men with initiative, education and skill were among the first to lay aside their tools and join the colors. This was a stroke of such folly that it took the country quite a long time to recover from it. Not only had we lost men of brain and initiative but also a large proportion of the skill for carrying on a war in which machinery and munitions were a preponderating factor.

When the sudden call came for an enormous increase in guns and ammunition the folly of the step that had been taken was realized and arrangements were made to effect the return of a number of skilled men from the colors. This, however, was difficult to carry out, and on top of it came the realization that if every skilled man were returned we had not sufficient to produce the munitions essential to success.

It was in these circumstances that the present Prime Minister, Mr. Lloyd George, then Minister of Munitions, laid down the principle that no man must perform any work that could be efficiently performed by women. He foresaw that every man would be required for essentially man's work. This involved the process which has become known as the Dilution of Labor. This dilution implies that:

(1) The employment of skilled men should be confined to work which cannot be efficiently performed by less skilled labor or by women.

(2) Semi-skilled and unskilled men should be employed on any work which does not necessitate the employment of skilled men and for which women are unsuitable.

(3) Women should be employed as far as practicable on all classes of work for which they are suitable.

This, at first sight, seems a simple arrangement, but I need not tell you that it is a complicated one and involves expenditure of time, patience and money on the part of the employers and sympathetic co-operation on the part of the skilled employees. Women have to receive careful training, intervening hands in the processes of production have to be upgraded and the skill in the factory has to be spread over a large area, usually necessitating more supervision and care if equal results are to be obtained. These are average conditions, but in a large number of cases the changes due to the introduction of women have resulted in considerably increased outputs over men's records, and this not only on light repetition work, but on heavy turning and laboring work and skilled and semi-skilled non-repetition work.

The patriotism shown by the employers in meeting the numerous difficulties which confronted them in training women for every conceivable kind of work, in reorganizing their factories for new productions by new labor, adapting their machinery to the measure of skill and strength available to work them has been one of the most inspiriting experiences of the war. Women have been trained in an incredibly short time, handling appliances of all kinds have been installed, special tools and gauges have had to be made and often when a work has been set fairly going and was reaching full production, a change in the country's needs for munitions compelled the Ministry to alter a design or a size or to put a firm on to a completely new product, and again the whole process of retaining employees and adapting tools and plant had to be gone through.

If the employer has done well for the country, so, on the other hand, no finer example of patriotism has been shown by any class than by the artisan as a whole. On an appeal being made to the Trades Unions soon after the outbreak of the war these Unions without a single exception agreed to do what the national interests required. A network of Trades Union rules and regulations, usages and customs, the result of many years of activity by organized labor, were freely set aside to allow for the introduction of women in nearly every class of work, subject to reasonable conditions.

Whatever some employers may think this was a gigantic sacrifice to make. It involved wages, hours, overtime, night work, Sunday work, meal times, holidays, shop regulations and demarcation arrangements, restriction of output, preparations of apprentices and classes of employees to be engaged, etc., etc. This structure of Trade Unionism the workingman agreed should be swept away to allow of the production of the maximum amount of munitions by the readiest methods and by any character of labor that was available—male or female—on the condition that the structure should be replaced at the close of the war.

This is splendid patriotism and when you add to the sacrifice of Trades Union Rules the burden, such as has fallen principally on the Trades Union men, of training the hundreds of thousands of women to do their work, this must increase our debt of appreciation and gratitude. It is only by the loyal co-operation of employer and employed that we are in the satisfactory position as regards munitions that we are to-day.

By the process of dilution we have been able to place in munition works about 950,000 women to do work from the heaviest laboring unskilled operation to the highest grade of toolroom non-repetition work. I do not hesitate to say that women have entirely destroyed our pre-war ideas as to what constitutes "skilled" work. When in the early days of the war women were trained to turn out 18 pdr.

H. E. shell and equal the production of male labor many thought that such work, amounting as it does to little more than manipulative dexterity, was about the limit of the capacity of women who had not received a regular course of Engineering training. After a few months' workshop experience, however, women are to-day building the greater part of one of the best High-Speed Engines in the country, each woman setting her own tools and work, and able to machine any piece of work that the tool she is on will take. Women are building guns, including the fine fitting work on the breech mechanism, and the cutting of large screw threads up to a shoulder. They are doing most of the work in some shops on three and one-half ton Army Lorries and will do practically the whole of it if the war lasts much longer, including chassis erection and testing. They are doing important work in marine engine building, turning connecting rods, propeller shaft liners and doing practically all in some cases of the marked-off drilling. The Aero Engine, as you well know, is a very fine piece of mechanism and at the outset was considered a tool room job throughout. In some shops women are to-day doing the greater part of the work turning on Centre Lathes to half a thousandth, milling webs of Clerget Cylinders on a booker Miller without stops and setting up their own jobs and working again to half a thousandth limit, boring cylinders on a No. 9 Herbert and similar work on a Gishlet, setting up their own jobs, turning and finishing test pieces in various metals to a 5,000th; making tools and gauges of all kinds to fine limits; all varieties of bench fitting to drawings and marking-off work of every description. Locomotive work, steel constructional work, boilers, bending, drilling and riveting. Women are doing magnificent work both in regard to accuracy and output.

On shells of all nature women should, of course, be principally employed. Contracts for shell will only be renewed and continued after March 31st next (1917) with those firms who employ 80 per cent. of female labor on shell of sizes from 2.75" to 4.5" inclusive. On larger sizes of shell, contracts will only be renewed if the Ministry of Munitions' instructions in regard to dilution have been carried out, not only in regard to the proportion of women to be employed in each factory but the proportions of semi-skilled men.

9 789361 380471

Printed by Libri Plureos GmbH in Hamburg,
Germany